Alkali-Activated Fly Ash Blast Furnace Slag Composites

Alkali-Activated Fly Ash Blast Furnace Slag Composites

Kushal Ghosh

Partha Ghosh

CRC Press
Taylor & Francis Group
Boca Raton London New York

CRC Press is an imprint of the
Taylor & Francis Group, an **informa** business

First edition published 2021
by CRC Press
6000 Broken Sound Parkway NW, Suite 300, Boca Raton, FL 33487-2742

and by CRC Press
2 Park Square, Milton Park, Abingdon, Oxon, OX14 4RN

CRC Press is an imprint of Taylor & Francis Group, LLC.

ISBN: 978-0-367-53554-4 (hbk)
ISBN: 978-1-003-08246-0 (ebk)

Typeset in Times
SPi Global, India

Dedicated to our parents

---

*They are, and always will be, our
greatest asset*

Contents

Preface

Alternate cement binder technology is an area of enhancing interest due to growing environmental concerns and considerable carbon footprint of the cement industries. CO_2 emissions can be reduced to a great extent by developing alternate binder like alkali-activated composites. They are an alternate sustainable green construction material for the 21st century from waste/low cost source materials such as fly ash and blast furnace slag, having much better performance than conventional cement composites. Alkali activation involves a number of processes such as dissolution, diffusion, polycondensation, and hardening. Optimization of such a complex system requires deep understanding of a number of synthesis parameters as well as of their interactions, which have been critically discussed. Second, the advantages of blending source materials such as fly ash and blast furnace slag have been explained to produce high-performance alkali-activated composites, indicating effects of various synthesis parameters and their relationship with mechanical properties and microstructure. Two materials of separate chemical characteristics are mixed together and subjected to alkali activation; the nature of the reaction products varies with the mix proportion; this fact has been explained in detail. A guideline has been indicated to have an optimal mix proportion which provides better overall performance, based on extensive experimental observations. Identification of the exact role of synthesis parameters and its relationship with the microstructure has also been highlighted. These issues have been discussed on paste and mortar through graphical and tabular form. Durability aspects have been discussed based on experimental observations in acid and sulphate solutions as well as exposing to different elevated temperatures up to 1000°C. It is observed that slag is more effective from strength and workability point of view whereas fly ash compensates its drawback contributing toward durability. The micro-structural, mechanical properties, and reaction products of alkali-activated fly ash–slag composites and their inter relationship are assessed and presented. A deep discussion on phase characterization has been made with the help of XRD analysis along with SEM. The change in the bond formulations of the atoms has been explained by FTIR, FESEM, and EDAX analysis. MIP results are used to understand the pore characteristics. A mix design guideline based on chemical compositions, relationship of synthesis parameters, target strength, workability, etc. (empirical statistical concept) has been put forwarded for practicing engineers to manufacture tailor-made alkali-activated fly ash composites in the presence of slag.

They are really a sustainable green material from waste/low cost source materials such as fly ash and blast furnace slag, having much better performance than conventional cement composites. They will replace conventional cement utilizing waste efficiently, promoting green technology. This book will be useful to researchers who are working at present and also to the teachers. Many intricate issues were discussed and resolved to a great extent. Manufacturers in the industry will be benefited to a great extent. All the important and intricate aspects were covered and given a holistic shape for the immediate use in the industry and in research.

Dr. Kushal Ghosh
Assistant Professor
Department of Civil Engineering
National Institute of Technology (NIT) – Sikkim – India

and

Dr. Partha Ghosh
Associate Professor
Department of Construction Engineering
Jadavpur University – Kolkata – India

Acknowledgments

It is our great pleasure to take this opportunity to extend our sincere thanks to Professor Dr. Mahesh Chandra Govil, Director, National Institute of Technology (NIT), Sikkim, India, for his continuous encouragement, support, and blessings.

We are greatly indebted to Head of the Construction Engineering Department, Jadavpur University, India for his co-operation. Our special thanks to all faculty members and staff members of the Concrete Laboratory and other laboratories of the Construction Engineering Department for their help and co-operation during our research work.

We would like to acknowledge the support Jadavpur University has extended through TEQIP and DST PURSE (Phase II) schemes of Govt. of India, by providing research facilities including fellowship.

We are grateful for the unending sacrifices of our family members and friends.

We are extremely thankful to Dr. Gagandeep Singh and Mr. Lakshay Gaba, of CRC Press – Taylor & Francis, for their guidance and help in writing this book.

Acknowledgements

Authors

Dr. Kushal Ghosh has received Bachelor of Civil Engineering with distinction from Pune University-India in the year 2012 and Doctor of Philosophy from Jadavpur University-India in the year 2018. At present, he is serving as Assistant Professor in National Institute of Technology (NIT)-Sikkim – India. He has worked as Structural design Engineer in Skematic Consultant – India for four years. His major research area is "Sustainable High Performance Green construction material". He has guided eight students for their thesis at ME level. He has published a number of papers in peer-reviewed journals and conferences mostly on "Alkali-Activated Composites" and on "High-Performance Concrete".

Dr. Partha Ghosh has received Bachelor of Civil Engineering from National Institute of Technology (NIT)-Agartala-India in the year 2001, Master of Civil Engineering and Doctor of Philosophy from Jadavpur University-India in the year 2003 and 2007, respectively. At present, he is serving as Associate Professor in Jadavpur University – India. His major research area is "High Performance construction material". He has guided fifteen students for their thesis at ME level and has guided two for their thesis at Ph.D level. He has published a number of papers in peer-reviewed journals and conferences mostly on "High Performance construction material". He has published two books out of which one book is based on "Geopolymer". He is a fellow of Institution of Engineers (India). He has rendered his services as Structural consultant at National level. He has done research in North Western University – USA, Monash University – Australia and Delft University-Netherland.

1 Introduction

1.1 INTRODUCTION TO ALKALI-ACTIVATED COMPOSITES AND HISTORICAL DEVELOPMENTS

Portland cement is extremely popular for manufacturing of concrete. It has advantages as well as disadvantages. It provides high strength, but it is not highly durable. Carbon dioxide emission during manufacture of Portland cement is also another problem [1]. Previously, Portland pozzolana cements are manufactured due to scarcity of limestone as well as for utilizing fly ash, slag, etc., which come out from different industries as waste. These wastes are used in blended ordinary Portland cement, which is quite popular. It reduces the use of limestone, and durability problems are solved to some extent. It is necessary to synthesize high-strength and highly durable concrete. In this context, alkali-activated composites are taking the place of conventional concrete. Alkali-activated composites are formed when an alkaline activator solution is mixed with an aluminosilicate/calcium-rich material such as fly ash, slag, etc., to produce different forms of alkali-activated composites through polymerization.

One of the main by-products of the thermal power industry using coal as a fuel source is fly ash. Fly ash is a material available in powdered form with a spherical shape and high amorphous content and is one of the major industrial by-products from thermal power plants. Fly ash is majorly composed of oxides of silica, alumina, iron, calcium, magnesium, sulfur, sodium, potassium, and also residues of carbon. The chemical composition of fly ash particles is greatly influenced by the source of the coal. Commonly four types of coals are used for power generation, namely, anthracite, bituminous, sub-bituminous, and lignite. Fly ashes with anthracite or bituminous coals as sources have oxides of silicon, aluminum, and iron comprising more than 70% and oxides of calcium being present anywhere between 1 and 12%. Lignite-sourced fly ash contains oxides of silicon, aluminum, and iron in the range of 50%–70%, and the content of oxides of calcium exceeds 20%. As per ASTM C618, fly ash can be classified into Class C, Class F, and Class N. Only Class F (CaO < 20%) and Class C (CaO > 20%) fly ash materials are used as prospective binder materials in the construction industry. As fly ash is produced at a high rate across the world, its storage and disposal pose a major logistical hurdle. A significant amount of fly ash produced is used currently, with majority of its use in the civil engineering sectors. Class C fly ash with a high amount of CaO can be used as a supplementary cementitious material, and upon reaction with alkaline pore water, it forms a secondary network of calcium silicate hydrate gel. Fly ash has also been known to be used as a material for stabilization of soil. It also serves as a good source material for manufacturing of alkali-activated composites mainly because of its spherical shape and high percentage of silica and alumina. Individual fly ash particles are composed of two types of solid spheres known as cenospheres (hollow spheres) and those

1

accommodating smaller spheres inside (plerospheres). When alkaline solutions come in contact with fly ash spheres, they start the process of dissolution of Al and Si ions from the surface of the ash into the alkaline solution. As the reaction progressed, the alkali ions infiltrated through both the cenosphere and plerosphere and then the dissolution of the amorphous portions of fly ash particles further continued to occur. Gradually, the plerospheres are filled with the dissolved ions as well as the activator solution as the activator solution continues to attack the sphere outwardly as well as inwardly. With time, the dissolved monomers react among themselves to start the polymerization process, which ultimately leads to the formation of polymer gels. An increase in the fineness of the fly ash particles leads to a higher reaction rate, which subsequently results in a lower setting time as well as higher mechanical strength. With regard to alkali activation, it has been seen that Class F fly ash is more reactive in nature as its amorphous content is comparatively higher. However as the energy required for the alkali activation reaction to progress is comparatively high, heat curing is required. For in-situ production of fly ash-based alkali-activated composites, this process of curing is a problem though ambient temperature is more suitable for applications.

Slag is a by-product created from the manufacturing process of different types of metals. Slag is composed mostly of oxides of alumina, silicates, calcium, and magnesium. Slag is widely used in the concrete industry as a supplementary cementitious material. The specific type of slag used for concreting is obtained from the steel and iron industries and is known as blast furnace slag. Blast furnace slag is generally formed by suddenly cooling the molten slag by immersing it in water. This action alters the molten slag into particles with size less than 4 mm. These particles are largely made of amorphous compounds as sudden cooling does not allow complete crystallization to take place. As slag is rich in calcium and aluminum, it demonstrates good binding properties on reaction with an alkaline medium. The behavior of slag depends on the raw materials being used to manufacture the required metal. The variation in raw materials affects the behavior of slag when subjected to alkali activation. Therefore, it is preferable to perform multiple trial mixes before using the slag for large-scale applications.

Commonly used activator solutions for the alkali activation of various source materials are sodium hydroxide (NaOH), potassium hydroxide (KOH), potassium silicate (K_2SiO_3), and sodium silicate (Na_2SiO_3). These activators have been chosen on the basis of their ability to dissolve aluminum, silica, and calcium ions from various raw materials such as fly ash and slag, which provides a favorable environment for the dissolved ions to converge into polymeric gels. Generally, higher alkaline concentration of the activator solution leads to a greater amount of dissolution. Sodium hydroxide has been seen to function more effectively than potassium hydroxide in liberating a greater amount of silicate and aluminum ions. This is because the size of sodium cations is smaller than that of potassium cations allowing them to smoothly navigate through the gel matrix and the polymerization process to occur. Sodium silicate is also widely used as an activator solution. Various other activators such as $Ca(OH)_2$, MgO, $Ba(OH)_2$, and $Sr(OH)_2$ have also been used as an activator. It has also been seen that multicompound activators tend to perform better than when they are used individually. Blended activator solutions have been found to significantly enhance the mechanical properties of alkali-activated composites. The most commonly used blended activator solution is a mixture of sodium hydroxide and sodium silicate solution. The composition and concentration of the activator solution determine important characteristics such as setting time, shrinkage, etc. Most of the activator solutions are corrosive in nature and present certain health hazards. The manufacturing process of these activators is also quite energy intensive in nature. Thus, it is imperative to manufacture alkali-activated composites with a very low concentration of activator solution, which will lower the health hazards as well as make the composite more sustainable in nature.

1.2 ADVANTAGES OF ALKALI-ACTIVATED FLY ASH–BLAST FURNACE SLAG COMPOSITES

Polymerization of fly ash-based alkali-activated composites occurs in four stages, namely, dissolution, gelation, solidification, polycondensation, and crystallization. These four stages may occur discreetly or simultaneously to form a solid material with superior strength and durability compared to those of Portland cement [2,3]. Fly ash and blast furnace slag are used as precursors in the manufacturing of alkali-activated composites [4–9]. Between the two materials, fly ash is the more popular choice because it is more abundantly available and its disposal is a problem. Safe and effective disposal of effluent, sludge, and other waste products such as fly ash formed because of coal combustion is a major challenge faced by the industries today. In the year 2015, the amount of fly ash produced in India was 176.74 million tons, and fly ash utilization was 107.77 million tons [10]. Most of the unutilized ash is disposed in landfills at suitable sites [11,12]. Landfilling is not a favorable option as it increases the total cost of running foundries and also has its own set of regulatory approvals . Ground water contamination due to toxic metals present in landfills is also of critical concern. Fly ash has reactive silica and also possesses good acid neutralization properties [13,14]. Fly ash has a comparatively low reactivity and needs a highly alkaline liquid for dissolution. However, studies performed with blended precursors have been seen to yield more favorable results in terms of strength, workability, and durability than composites manufactured with sole precursors [15,16] . Addition of slag, a material already being used in the construction industry in various forms [17–19] , has been seen to vastly improve the reactivity of the alkali-activated mix leading to better workability and high early strength. The polymerization of fly ash requires thermal curing of about 70–85°C to form suitable alumina-silicate reaction products [19,20]. Ismail et al. [21] and J Oh et al. [22] have reported the reaction products of the alkali-activated fly ash–slag composite. Addition of alkaline solution to slag leads to the disintegration of calcium and formation of aluminum precipitates which leads to the formation of a C-A-S-H type gel [21–23]. This gel functions as the major binding agent contributing to the increased level of mechanical strength, decreased level of porosity, and high durability. Polymerization of fly ash leads to the formation of a cation-activated alumina-silicate gel, that is, a sodium alumina-silicate gel N-A-S-H [23]. Alkali activation of a blended precursor containing slag generally leads to the formation of a hybrid binder system comprising both C-A-S-H and N-A-S-H gels. The nature of the cross-linkage has been found to depend upon the chemical composition of the precursors and the concentration of the alkaline activator solution. Lloyd [24] suggested that addition of slag produced C-S-H-type reaction products with interlinked aluminum ions depending upon its availability. The impact of slag substitution on the fly ash-based alkali-activated composite and its resultant effect on reaction kinetics was studied by Kumar et al. [23], and they found that the C–S–H gel forms a major portion of the reaction products along with alumina–silicate gels. This concurrent existence of both the gels leads to high early settings and better mechanical properties [23–26] . Provis et al. [15] via X-ray microtomography data also established the dominant presence of slag-based reaction products at a microstructural level. A study conducted on the microstructural aspects of younger-age pastes by Yang et al. [27] also established the fact that addition of slag into sodium hydroxide-activated fly ash has an effect on the reaction mechanism, which is one of the major controlling factors behind the formation of the main binding gel products. Most of the previous studies discussed above used water curing at ambient temperature as slag particles do not need heat curing to reach the hardening stage of alkali-activated binder systems. But to enable the fly ash to participate fully in the polymerization process, curing at a higher temperature of 70–90°C is needed [20,21]. It has also been reported that the fire and acid resistance [28–31] of fly ash-based alkali-activated composites is better than that of slag-based alkali-activated composites. Thus to harness the true potential of a fly ash-based blended alkali in the presence of slag, one has to ascertain mixing and curing conditions that allow the source materials to fully take part in the polymerization reaction.

Extensive experimental investigation was carried out on blended (fly ash + slag) alkali-activated composites by the author, in both paste and mortar. It was observed that the hydroxide content, silicate content, and water/binder ratio have a significant effect on the fresh and hardened properties of fly ash-based blended composites. The dissolution rate of the fly ash particles depends on the concentration of sodium hydroxide and sodium silicate used in the activator solution. The amount of water also played a crucial role in determining the quality of casting and level of compaction that was needed. Though fly ash-based alkali-activated composites have exhibited superior performance, they have problems of low reactivity and workability in a pure alkaline environment. In order to increase the overall reactivity, fly ash was blended with ground granulated blast furnace slag to have a more reactive and improved blended alkali-activated composite. Alkali activation of the above-mentioned mix undergoes different reaction mechanisms and leads to the creation of separate reaction products which are distinctly different from the composite where only fly ash or only slag is used. Alkali activation of fly ash undergoes polymerization leading to the formation of an aluminosilicate hydrate gel product, whose gel structure is three dimensionally tetrahedral in nature, whereas when slag reacts with an alkaline solution, the binder gel is formed, which has resemblance to the calcium silicate hydrate gel formed during hydration of ordinary Portland cement. Thus, when these two materials of separate chemical characteristics are mixed together, subjected to alkali activation, the nature of the reaction products will vary from products having only fly ash or only slag.

Blended alkali-activated composites were investigated extensively, and results are presented in chapters 3, 4, and 5. It was observed that the mechanical properties and durability characteristics as well as the flow/workability of the blended alkali-activated composites improved remarkably compared to products with only fly ash or only slag.

1.3 APPLICATIONS

Alkali-activated composites have been used extensively in various industries through numerous commercial applications. Notable examples of commercial applications are blocks, pavements, pipes, and sinks made from alkali-activated concrete. They have also been used in structural applications. Manufacturing of autoclaved aerated concrete and refractory concrete has also been performed with alkali-activated binder systems. Earlier, countries such as the formerly known USSR and China and in recent times, Australia, UK, and USA have used alkali-activated binder systems successfully. Alkali-activated systems have been seen to provide high strength and durability characteristics such as superior resistance to fire, sulphate, and acid attack, respectively. In addition to the abovementioned advantages, the alkali-activated binder system demonstrates the following properties such as better adhesive strength, surface hardness, and low level of macro- and micro-level porosity. Because of these properties, alkali-activated binder systems are used to manufacture components designated for military use, railway sleepers, etc., and as a result, alkali-activated composites are being increasingly being looked upon as an alternative to ordinary Portland cement-based concrete. As far as source materials are considered, there are a variety of industrial wastes such as fly ash, blast furnace slag, rice husk ash, red mud, waste paper sludge, etc., which can be used. This serves the dual purpose of manufacturing an alternative sustainable green binder system and a proper utilization of waste materials.

1.4 FUTURE PROSPECT AS A GREEN SUSTAINABLE MATERIAL

The manufacturing process of ordinary Portland cement entails a huge amount of energy consumption which can be specifically attributed to the calcination of the raw material performed at a temperature of 1400°C and 1600°C. Through this process, a mammoth amount of CO_2 is also released into the atmosphere. It has been estimated that the consumption of concrete is going to increase by 18 billion tonnes per annum by 2050 [32]. Cement production has increased substantially in recent times in countries in the Middle East and Northern Africa as well as in India and China. China leads the pact in terms of cement production with a market share of 58.13% [33]. Currently, 3.5 billion tonnes of cement are being produced with approximately 900 kg of CO_2 being released for every 1000 kg of cement produced. Furthermore in addition to CO_2, gases such as sulfur dioxide and nitrous oxide are also released which magnifies the greenhouse effect [34,35]. For the production of 1 tonne of cement, one needs 1.5 tonne of raw materials which are majorly procured from natural resources [36,37]. Thus, the cement industry increasingly wants to adopt a more sustainable environment friendly approach. Thus in this context, alkali-activated binder systems are gaining a lot of traction currently and are considered as an alternative sustainable binder system. Alkali-activated composites use industrial waste as a raw material, thus saving natural resources as well as providing an effective waste management system. The consumption of energy in alkali-activated binder systems is generated from the manufacturing process of the alkaline activator solution and the curing process. These problems can be mitigated by using blended raw materials with greater reactivity for the production of alkali-activated composites. This will help reduce the concentration of activator solution as well as the required curing temperature. Studies by Duxson et al. [38] have shown that CO_2 emission can be reduced by 80% if alkali-activated binders are used [39]. However, other research groups such as McLellan et al. have reported that alkali-activated composites can be produced at a cost of 9% lower to 39% higher than that of OPC. The contrast in the reports shows that more extensive evaluation is needed with regard to the life cycle of alkali-activated products, which will enable them to be used widely in mass commercial applications.

2 Polymerization and Microstructure of Alkali-Activated Composites

2.1 PREAMBLE

Alkali-activated composites involving fly ash are composed of aluminosilicate hydrate and have a polymeric structure. They are formed when an aluminosilicate-rich material reacts with an activator solution with medium to high alkalinity. This phenomenon leads to the formation of a three-dimensional tetrahedral polymeric structure. The resultant spatial arrangement created among the sialates tends to make the structure amorphous. Different types of source materials, for example, fly ash, metakaolin, blast furnace slag, etc., can be used to manufacture alkali-activated composites of varying specifications. The phase change mechanism during the polymerization reaction consumes a high amount of energy. This energy is generally provided through heat curing which acts as hindrance in the classification of alkali-activated composites as a sustainable material. Thus, a thorough understanding of the polymerization process is required to improve the performance of alkali-activated composites while simultaneously reducing the need for heat curing, to make them greener.

Polymerization of the abovementioned alkali-activated composites is an extremely complex process, and to understand the sequence of events leading to the phase change of the aluminosilicate source material into a polymeric binder material, a host of different experimental techniques are applied. A nanoscale explanation of the polymerization process may be provided by infrared spectroscopy and X-ray diffraction (XRD) analysis. Scanning electron microscopy (SEM) can also be used to gain an insight into the actual structural morphology of the polymeric products. The macroscopic changes complementing the polymerization process can be studied using calorimetry and rheology. Based upon the information collected through the abovementioned techniques, mathematical modeling can be performed to express the progression of the reactions and to factor in the different effects of various factors such as concentration of the activator solution, degree of amorphousness of the source material, curing temperature, etc. To fully understand the process of alkali activation, a detailed study of the reaction kinetics, degree of polymerization, role of non-evaporable water, pore solution chemistry, and silicate polymerization needs to be carried out. Such an understanding will lead to the development of tailor-made manufacturing processes providing greater control over the quality and nature of the alkali-activated composite products formed.

2.2 POLYMERIZATION

Molecules are entities made of different types of atoms. Water which consists of two molecules of hydrogen and one molecule of oxygen is an example of a small molecule. In comparison to water, polymers are very large in size. Polymers contain both amorphous and crystalline materials. Polymers have been observed to demonstrate superior engineering properties such as being lightweight, strong, and stiff simultaneously. They also possess higher toughness, resiliency, and corrosion resistance and lesser value of thermal conductivity as compared to non-polymeric materials. Molecular weight and crystallinity are some of the important factors controlling the properties of a polymer. Greater molecular weight leads to higher chemical resistivity up to a threshold value. It ensures that the decrease in strength occurs over a longer period of time as heavier molecules can sustain more damage. However, a higher molecular weight also leads to an increase in viscosity, which can be detrimental to having a smooth manufacturing process. The long-range ordering of structure is clearly demonstrated by its degree of crystallinity. A material with a greater amount of crystallinity tends to have more regularly aligned chains, resulting in the increase of hardness and density.

As compared to the hydration process of ordinary Portland cement (OPC) in which calcium silicate (C_2S and C_3S) is converted into calcium silicate hydrate (C-S-H) and calcium hydroxide $Ca(OH)_2$, hardening of alkali-activated polymers takes place through a process of polycondensation of oligomer- (sialate-siloxo) into a polymer (sialate-siloxo)-based cross-linked network. Alkali-activated polymers can be broadly defined as a combination of compounds comprising recurrent units, such as silico-oxide (-Si-O-Si-O-), silico-aluminate (-Si-O-Al-O-), ferro-silico-aluminate (-Fe-O-Si-O-Al-O-), or alumino-phosphate (-Al-O-P-O-), generated via the procedure of polymerization. Normally, sodium hydroxide or potassium hydroxide is used as an activator solution in the manufacturing of such composites. As there is a difference in the size of the two different cations with K^+ being larger compared to Na^+, it causes a slight change in the reaction kinetics of alkali activation. The Si-O-Al gel network of fly ash-based alkali-activated composites is quite comparable to that of zeolites. The polymeric gel network of alkali-activated composites is more amorphous than the structure of zeolites. Fly ash-based-alkali-activated composites when observed on a nanometer scale have a microstructure comprising minute aluminosilicate clusters of roughly 5–10 nm. These clusters are dispersed within a highly porous network. The reaction leads to the formation of poly(sialates) or poly(sialate-siloxo) or poly(sialate-disiloxo) by linking SiO_4 and AlO_4 into tetrahedral structural frameworks, the exact nature of which depends on the ratio of SiO_2/Al_2O_3. These tetrahedral frameworks are formed through covalent bonds which are long range in nature. Due to this, fly ash-based-alkali-activated composites have been broadly perceived to possess a three-dimensional aluminosilicate gel structure. Normally, alkali-activated composites are manufactured as a two-part mix comprising an alkaline solution and an aluminosilicate-rich source material. The polymerization reaction may take place at ambient temperature, but at a slightly elevated temperature, its intensity is more. The reaction causes the leaching of Al and Si ions into the alkaline solution which is followed by nucleation and condensation allowing the gel binder to transform into a solid binder. Though fly ash-based alkali-activated composites have mostly amorphous gel structures, they can be changed into crystalline ceramic phases such as pollucite and leucite upon extended period of heating. These polymers have a more rigid gel structure than normal amorphous alkali-activated composites.

2.2.1 Degree of Polymerization

The degree of polymerization influences greatly the final product characteristics with respect to strength and durability and can be viewed as a good indicator of product quality. It has been found to be directly related to the reaction mechanism and reaction kinetics. The operating conditions in the case of alkali-activated composites are curing conditions, concentration of alkaline activator solution, etc. and have a great degree of influence on the polymerization process, and minute changes in any of them will have a profound effect on product quality. The degree and distribution of the degree of polymerization depend on the microkinetics of the governing reaction. All the factors that can influence temperature and concentration distribution in the reactor, such as mixing, flow, mass transfer, and heat transfer, may have a great impact on product quality. Therefore, the key to properly understanding the polymerization process lies in analyzing reaction kinetics and transfer processes. Polymerization and in particular the degree of polymerization have an impact on many of its properties. The affected properties are tensile strength, impact strength, resistance to cracking, and viscosity.

2.2.2 Non-evaporable Water

Water plays a very important role in the polymerization of alkali-activated composites. It simultaneously acts as a medium of dissolution as well as a transportation medium for silicate and aluminate ions. Water is needed in the dissolution process, to foster the polymerization process and form monomers such as $SiO(OH)_2$, $Si(OH)_4$, and $Al(OH)_4$, or oligomeric species. It is also needed to initiate the polycondensation process as well [40,41]. A certain amount of water remains enclosed inside the polymeric gel network even after the reaction has taken place. Duxson et al. [42] Other research studies such as those by Dimas et al. [43] suggested that silanol groups contain non-evaporable water. This non-evaporable water remains in the gel network even after application of temperature of more than 1000°C. The non-evaporable water also abets the development of compressive strength at a later stage by making itself available for further dissolution of unreacted Al^{3+} and Si^{4+} compounds. This water also enhances the volume stability of the polymeric gel by providing reasonable humidity conditions [44].

The effect of water on the polymerization process of alkali-activated composites is still not fully understood. Water is present in different states in alkali-activated composites, and it is important to distinguish between them on the basis of their origin. A conceptual model needs to be established, which accounts for the various states of water within the gel structure as it plays an important role in the polymerization process. Studies involving XRD and thermogravimetric analysis demonstrated that the presence of excess residual water hindered the development of stable crystalline phases. Calorimetric studies have shown that a high liquid/solid solution increases the rate of dissolution, but it slows down the stage involving polycondensation. Non-evaporable water has been found to have a stabilizing effect on the gel structure, and there is a threshold value. The amount of non-evaporable water depends on the amount and type of hydration products. Similarity can be found with pure cement samples where the amount of non-evaporable water is connected to the quantity of C-(A)-S-H gels. In the case of alkali-activated polymer composites, the gels that are formed are structurally disordered, cross-linked, and three-dimensional in nature, and it is difficult to establish the relationship between non-evaporable water and gel formation.

2.2.3 SILICATE POLYMERIZATION

The percentage of silicate content in the activating solution has been found to affect the polymerization at an early stage. Monomers, dimers, and trimers of silicates act as main species. The content of dimers and trimers increases as the silicate percentage increases [45,46]. Dimers have been seen to enhance the condensation stage of the polymerization. On the other hand, the presence of cyclic trimers decreases the rate of polymerization. Again, if the silicate and aluminosilicate are of smaller size, it leads to rearrangement of the gel and causes a higher degree of crystallization in comparison to mixes having a greater percentage of cyclic species of higher size [46]. Thus, it can be seen that the silicate polymerization greatly affects the reaction kinetics and subsequently the various stages of polymerization. It has been seen to greatly affect strength as well as porosity at a microstructural level [47,]. During the reaction, it has been observed that an Al-rich gel is formed first followed by a silica-rich gel [48], though the latter is difficult to identify through spectroscopic techniques as the silicates and aluminosilicates may display similar characteristics [49]. The polymerization of silicates occurs at a comparatively low pH, as a higher pH(>14) causes a level of depolymerization in the silicate network [50–52]. The Q numbers of a silicate solution also influence the reaction as they have a controlling effect on non-bridging oxygen (NBO) atoms which in turn affects the relative distribution of SiQ_n (n = 0–4) units in the solution. In case silicate solutions have a low Q unit, the stable aluminosilicate takes a long time to participate in the gelation process.

2.2.4 KINETICS OF ALKALI-ACTIVATED POLYMERIZATION

Reaction kinetics is a valuable tool for understanding the reaction mechanisms accurately. One of the major reasons behind studying kinetics is that it helps to ascertain the time period needed for a particular combination of reactants to reach the equilibrium state. Reaction mechanisms refer to two things in general. The first concerns the overall summarization of the steps involved in the reaction, and the second refers to the details of the individual steps themselves. There are also different forms of kinetics such as chemical kinetics which can be defined as the study of chemical reaction rates, and the different molecular processes involved in the reactions occur where transport is limited. On the other hand, kinetics pertaining to nucleation is defined as the arbitrary formation of a thermodynamically well-defined new phase (daughter phase or nucleus (an ensemble of atoms)) that will have the potential to irreversibly transform into a bigger sized nucleus inside the body of a metastable parent phase.

The rate of reaction and its changes with regard to synthesis factors such as curing temperature, chemical composition (Si/Al and Na/Al ratio), and concentration and composition of multi-compound alkaline activator solution are commonly studied with the help of an isothermal conduction calorimeter (ICC). Simulation of reaction mechanisms using molecular dynamic modeling has made the understanding easier with respect to the rate of reaction, and the investigation of reaction mechanisms with molecular modeling has shed light on the realizable reaction pathways and oligomers for polymerization of polymer gels.

There have been quite a few research studies to assess the effect of alkali concentration on the reaction kinetics and mechanism of fly ash polymerization. The parameters controlling the kinetics of the polymerization with respect to different alkali solutions can be understood by the usage of data regarding the emission of heat during reaction. These data can be measured by using an ICC, to monitor the reaction at different temperatures. When the alkali concentration increases, the calorimetric peaks get intensified, and there is an increase in the quantity of the total heat that is emitted. As the alkali concentration increases, there is a higher amount of dissolution of Al and Si which raises the requirement of energy needed for the reaction to proceed. The required minimum activation energy can be evaluated by calorimetry. The polymerization can also be divided into three stages, namely, dissolution, gel formation, and restructuring. In general, it was observed that the fly ash polymerization is kinetically favored in between 60 °C and 85 °C temperature for different alkali concentrations. The particle size of the source materials also has an effect on the kinetics of the reaction. The rate of reaction increases if there is a decrease in the average particle size. Studies have been conducted in this regard where the fly ash particles have been milled before usage. The apparent activation energy of the reaction lessened with decrease in the particle size as finer fractions are more susceptible to alkali activation. Though changes in the particle fineness cause an alteration in the kinetics, the reaction mechanism which is controlled by nucleation and growth remains unchanged. The activation energy required around ~100 kJ mol^{-1} [53].

The chemical composition of the binder material and the ratio of the binder material to activator concentration have a significant influence on the reaction kinetics of alkali activation. The effect of the chemical composition of the binder (slag, fly ash, or both), varying degrees of alkalinity, is generally shown using the ratios of Na_2O-to-binders (n) and activator SiO_2-to-Na_2O ratios (Ms). The effect of sodium silicate solution on the properties of fly ash–slag blended systems is extremely significant. The binder composition and the n values are chosen on the basis of the setting times required. A greater amount of slag percentage in the binder material leads to the reduction of the required n value.

The polymerization model of fly ash-based alkali-activated composites can be broken down into multiple stages. The universally accepted stages are dissolution, precipitation, reorganization, gelation, and polycondensation. Duxon et al. [54] have suggested a simplified model to explain the reaction mechanism. In the first stage, the aluminosilicate raw material undergoes dissolution through hydrolysis, and it creates aluminate and silicate species. The chemical composition and structure of the raw materials are major factors affecting the dissolution stage of polymerization. In the stage concerning reorganization, the Al^{3+} and Si^{4+} species after dissolving are converted into oligomers, and this stage is followed by condensation in which comparatively large networks are created leading to the production of the polymeric gel. This gel structure continues to undergo evolution by restructuring and rearranging into an amorphous part crystalline three-dimensional aluminosilicate gel. Historically, aluminosilicate-rich raw materials are used as polymer precursors. Among the available raw materials, fly ash, a powder-based raw by-product, is generated from thermal power plants which produce fuel by combustion of coal. The advantages of fly ash are its straightforward availability, low price, spherical shape, and presence of amorphous aluminosilicate-rich compounds. During polymerization of fly ash, it is seen that the initial reaction rate is high but it slows down with time. The polymerization of fly ash is also controlled by its chemistry, mineralogy, and morphology as well external conditions such as alkali concentration, reaction temperature, curing temperature and duration, etc. These parameters can be best judged by detailed calorimetric studies in which the abovementioned parameters are changed and the variation in heat emission is recorded.

Provis et al. [54] have also tried to summarize this process in four simple steps:

1. The alkaline ions attack the aluminosilicate source and cause the release of Al and Si species into the solution. For a metakaolin-based system, in the case of the metakaolin system, five- and six-coordinated Al is transformed into four-coordinated aluminate species when subjected to dissolution.
2. The newly dissolved species and the silicates made available through the activator solution are transported via the water available in the mix, allowing them to interact. These interactions lead to the formation of Al–Si oligomers.
3. The dissolved species starts to precipitate into an amorphous gel when the silicate solution is destabilized by the existence of dissolved aluminate species in high concentration. The dissolution and gelation processes occur concurrently, and they may also be influenced by additional factors such as the existence of reactive aggregates or oxides functioning as nucleation sites or mechanical disruption such as shearing or ultrasonication.
4. Growth of an amorphous gel with a cross-linked structure three-dimensional in nature continues until the reacting slurry solidifies. The time taken for the formation of the structure depends on the mix design along with the curing temperature and presence of contaminants. Depending on the above factors, the setting time can vary from a few hours to a number of days. The reaction process continues for a long time period after the setting has occurred and bears similarity to the growth of zeolitic crystallites.

The reaction kinetics of fly ash-based polymers is still not fully understood due to reasons such as the heterogeneous nature of fly ash, irregular nature of the reaction between fly ash and alkali, existence of impurities in the reaction system leading to the generation of complex reaction pathways and subsequent alteration in the reaction mechanism. The presence of unreacted fly ash and alkalis in the system also causes difficulties in the characterization of the gel products.

In order to tailor the properties of alkali-activated composites to suit specific needs and applications, understanding the relationship between reaction kinetics, microstructure, and mechanical properties of alkali-activated composites is needed. For this purpose, multiple experimental and computational studies have been performed by various research groups. Analytical techniques such as XRD and SEM-EDX are very helpful in the predicting the chemical composition and microstructure of the gel. The changes at the molecular level can be better understood with techniques such as ATR-FTIR, NMR, and in-situ X-ray/neutron PDF. ATR-FTIR (Attenuated Total Reflectance-Fourier Transform Infrared Spectroscopy) is an effective analytical technique by which the changes in the reaction at the molecular level can be understood. It is a technique based on both classical and quantum vibrational theory, through which different types of chemical bonds can be identified through the absorption of infrared radiation, allowing the identification of different chemical bonds in a molecule from the absorption of infrared radiation at different wavelengths. Materials are characterized according to the normal modes of vibration which corresponds to a change in the dipole moment of a molecule. The vibrations may constitute stretching which leads to a variation in the interatomic distances and also results in bending and other deformations. The analysis of FTIR spectra is quite complex as bands are made of fundamental vibrations as well as a combination of overtone vibrations. The analysis is performed by correlating the chemical structure of compounds with specific spectral features. The FTIR technique is a standard well-accepted method. A review of the FTIR literature related to silicates, aluminates, and aluminosilicates may be discussed. The fundamental theory of the attenuated total reflectance (ATR) technique is discussed, and the experimental apparatus and procedures for data collection are outlined. The ATR technique is quite different from conventional transmission infrared spectroscopy as it allows absorption spectra to be calculated for opaque samples, without dispersing in a transmitting matrix. In conventional transmission FTIR, a minute amount (0.5 mg) of the sample is ground with roughly 5 mg of potassium bromide and compressed into a disk. As fly ash is not at all homogeneous, a greater amount of sample volume is preferred. The grinding process may also cause an alteration in the structure of a material and cause a deviation in the results. Another issue regarding recording FTIR spectra on fly ash-based polymers is that potassium bromide dissolves in a silicate solution. Now silicate solutions may be present in the pores of the mix in the early stages of gel formation. To mitigate this issue, the samples are dried which again leads to precipitation causing a change in the structure and ultimately in the reaction kinetics. Thus, in order to perform a more accurate fly ash-based alkali-activated spectral analysis, another technique called ATR-FTIR is used. By using ATR-FTIR, one can calculate the transmission spectra of solids, gels, and solutions concurrently. The sample needs not be destroyed in the process.

A significant amount of literature is there on FTIR analysis of fly ash-based alkali-activated composites. However, the reported results are quite conflicting in nature. Still the basic FTIR spectrum peaks of amorphous silica taken at 1100, 800 and 475 cm^{-1} can be linked to three main bands which were associated with the stretching, bending, and rocking phenomenon of the Si-O-Si bonds, respectively. Previous research suggests that analysis of the activated fly ash pastes shows the influence of heat curing conditions on the FTIR spectra of fly ash mixes. The stretching vibrations related to Si-O-Si bonds of mixes which have been heat cured for 48 hours show wavenumbers of a higher order in comparison to mixes subjected to ambient curing. This is mainly due to the change in Si/Al ratio in the subsequent reaction product. FTIR spectral patterns can be used to predict the nature of reaction products formed. In the case of alkali-activated composites, the FTIR patterns indicate that the microstructures of the alkali-activated composites are of disordered nature. Most of the bonds appear in the FTIR patterns in terms of broad bands or shoulder, instead of sharp peaks, which confirmed the disordered microstructure of geopolymer gels. The most well-defined band of the polymeric gel may be said to be the Si-O-T (T=Si or Al) asymmetric vibration band. This band is found to be in the range of 900–1300 cm^{-1}, which is usually assigned to the main band of geopolymer gels. The wavenumbers of the FTIR spectra of the raw materials are always less than the wavenumbers of the polymeric reaction product. This helps in comparing the spectra of the raw material and the polymeric product to understand the formation of the polymeric gel. The wavenumbers representing different polymeric bands are provided in Table 2.1.

2.2.5 Pore Solution Chemistry

Pores form an integral part of the microstructure of alkali-activated polymer composites. They cannot be called a part of the polymer gel structure but they are formed during the polymerization process. The porosity of the polymeric gel has a significant effect on the mechanical properties of the alkali-activated composites. It also directly controls the permeability of the composites, thus affecting the durability in a major manner. The volume of pores and their size distribution can be assessed by mercury intrusion porosimetry (MIP), N2 adsorption-Barrett–Joyner–Halenda (BJH) analysis, and microcomputed tomography (microCT) techniques. The pore volume and pore size distribution of polymers have been assessed by MIP, N2 adsorption-BJH analysis, and microCT techniques. Depending on the connectivity, the pores can be categorized as open pores and closed pores, the latter term referring to the pores which

---

TABLE 2.1

FTIR bands assigned to different types of alkali-activated composites (Panias et al. [86])

Assignment	Wavenumber (cm^{-1})
Stretching vibration (–OH, HOH)	3500–2300
Bending vibrations (HOH)	1650–1630
Stretching vibration (O–C–O)	1430–1410
Asymmetric stretching vibration (T–O–Si, T Si, or Al)	1090–990
Si–O stretching, OH bending (Si–OH)	882
Symmetric stretching vibration (Si–O–Si)	800–780
Symmetric stretching vibration (Si–O–Si and Al–O–Si)	550–750
Bending vibration (Si–O–Si and O–Si–O)	460–470

cannot be measured by MIP or N2 adsorption-BJH analysis. The pores may also be categorized based on diameters. Pores with sizes of a few nm to 0.02 µm and 0.1 to 1 µm or larger are classified as gel pores and capillary pores. Generally, the pore size is believed to lie in the range of less than 10 nm to approximate 10 µm though the pore size distribution greatly varies with the chemical nature and its synthesis parameters. The range of the measured pore size depends on the type of technique being used for evaluation. The pore sizes for fly ash-based alkali-activated polymers are in the range of 0.5–5 µm and 10–50 µm, while the general pore size using the N2 adsorption–BJH analysis is in the range of 1.7–8 nm and 10–100 nm. The inter-connectivity, shape, size, and tortuosity of pores in alkali-activated polymer composites can be analyzed and visualized in a better manner by using microCT technology.

The solution entrapped in the pores serves as a medium of dissolution for the different types of aluminosilicate and calcium-rich precursors. The composition of the pore solution is a good indicator of the stage of the reaction. If the pore solution has a high concentration of hydroxyl ions, it will cause an increase in dissolution by expediting the fragmentation of Si-O and Al-O bonds [55–59]. As dissolution directly affects the subsequent stages of polymerization such as reorganization, gelation, and polycondensation, the pore solution chemistry influences the reaction kinetics as well as the nature of the final reaction products. The alkaline activator dissolves the precursors to form a saturated or an oversaturated solution and controls the reaction mechanism in the process [60,61]. If the pore solution is adequately alkaline in nature, then it ensures the stability of the polymeric reaction products, which in the case of alkali-activated fly ash slag composites are alkali calcium-aluminosilicate hydrate (C-(N-)A-S-H) type gel or a three-dimensional hydrous alkali-aluminosili-cate (N-A-S-H) type gel, which depends on the amount of Ca present in the system [62,63]. If the composites are embedded with reinforcements, then the pore solution also controls the potential for susceptibility of the reinforcements to corrosion [64,65]. The type of activator solution also influences the pore solution chemistry. When only sodium hydroxide is used to activate the fly ash, the dissolvable Al from the fly ash creates a silica-rich layer on the surface and stops the hydroxyl ions from coming in contact with the unreacted portion of the fly ash. When soluble silicate is also used in the activator solution, it can prevent the Al from being absorbed to the surface and allows quicker dissolution of the fly ash particle [66]. The nature of the source material also has an effect on pore solution chemistry. As per a study by Zuo et al. [67], the concentration of Si and Na was 37.5 mmol/L and 1670.4 mmol/L for a blended fly ash and slag paste and 11.7 mmol/L and 2517.7 mmol/L for a slag paste. The concentration of Si for the fly ash slag paste is more because it has a higher content of reactive silica. Proper understanding of the volume and distribution of pores and pore solution chemistry is extremely crucial for understanding the relationship between synthesis factors and mechanical and microstructural properties of alkali-activated polymer composites.

2.2.6 POLYMER GELS

Through the process of polymerization of fly ash-based alkali-activated composites, which is more popularly known as geopolymerization, the polymeric gels are formed.

The nature of the polymeric gel mainly depends on the chemical composition of the source material and the activator solution. This gel acts as the primary binding agent in alkali-activated composites and controls the mechanical as well as microstructural properties. Even though extensive experimental and numerical studies have been conducted, the exact structure of the polymeric gel is not yet understood. Various sophisticated analytical techniques have been used to understand the nature of the gel such as FTIR, NMR, XRD, and SEM to arrive at a conclusion.

The general understanding is that in the initial stages of the polymerization reaction, an Al-rich gel is formed which is later enhanced with silica. A tetrahedral polymeric network is formed where the negative charges of Al and Si are balanced with the positive cation of the hydroxide solution. The network also consists of NBO atoms. If the amount of Al^{3+} increases, it lengthens the bonds and significantly weakens the bonds to adjacent Si-O groups. It has been observed that the reaction mechanism of alkali-activated fly ash composites has a profound similarity to zeolite synthesis. At first, Al from the fly ash starts to leach producing an Al-rich gel. This leads to a creation of a fly ash surface lacking aluminum and also consisting of a partially disconnected silicate network. This layer further dissolves allowing the release of Al and Si species from the fly ash. With further passage of time, there is a continued diffusion which makes the aqueous solution more homogeneous in nature. Slowly, the gel achieves a state of quasi-equilibrium through multiple stages of depolymerization and re-polymerization reactions. The effect of additives such as slags also has an impact on the gel formation. It is seen that the dissolved species of slag present in the solution has an effect on the dissolution of the fly ash particles and subsequently on the rate of fly ash–slag–based alkali-activated composite gel network formation. The effect of slag on the gel structure has been discussed in detail in Chapter III.

2.2.7 SELECTIVITY OF ACTIVATORS

Alkaline hydroxides and silicate solutions are the most widely used activating solutions. Alkaline hydroxides are formed from an alkali metal cation and a hydroxide ion. Among alkali hydroxides, sodium and potassium hydroxides are the most widely used. Between the two, NaOH being cheaper in comparison to KOH is used more. Using a greater concentration of sodium or potassium hydroxide results in the formation of polymeric gel structures resembling zeolites after a reasonable period of water or heat curing. Special precautions should be taken during mixing the activator solution and source material as heat is evolved during the reaction. In recent times, it has been seen that using multi-compound activator solutions gives better results in comparison to using only alkali hydroxide solution. In this multi-compound activator generally, a sodium silicate solution is used with alkali hydroxide solution. Sodium silicate is the general name for a chain of compounds with the formula $Na_2O \cdot nSiO_2$. The properties of sodium silicate may change depending on the number n. Liquefied sodium silicates are commercially available as waterglass. Different types of sodium silicate are available based on their silica modulus (SiO_2-to-Na_2O (or K_2O) ratio), which varies from 1.6 to 3.3. Commercially available waterglass tends to have a silica modulus of 1.60 to 3.85 and has a solid ratio of 0.36–0.4. Waterglass has been reported to be the one of the most potent activators for most alkali-activated composites [68]. Addition of silicate solutions increases the viscosity of the activator solution and subsequently of the mix as a whole. Potassium silicate can also be used, and it also has a lower viscosity as compared to sodium silicate but it is more expensive as well has larger size of cations, thereby making it less popular.

2.3 STATUS OF RECENT RESEARCH WORK

Geopolymers have immense potential to be used as a popular alternative construction material. The properties of geopolymers depend on different factors such as synthesis parameters, choice of source materials, etc. Researchers have published their findings reporting the superior strength and durability of geopolymers. Most of the studies dealt with single source-based geopolymers such as fly ash, metakaolin, etc.. It has been seen that the mixing proportion is the most important aspect in this regard to obtain a tailor-made geopolymer. Fly ash and blast furnace slag are the most widely used precursors in the manufacture of alkali-activated composites with a number of studies being conducted [4–9]. Between the two materials, fly ash is the more popular choice because it is more abundantly available and its disposal is a big problem. However, fly ash has the problem of low reactivity and slag has been reported to have low durability in comparison to fly ash. Thus, in order to make a better geopolymer, a source material with high reactivity in alkaline medium such as slag needs to be blended with fly ash. In order to obtain a blended geopolymer composite, it is extremely important to understand the reaction mechanism of alkali with source materials. Thus, a critical evaluation of the existing literature on fly ash-based geopolymers and alkali-activated slag (AAS) binders is presented in this chapter. It covers past to recent understanding of the reaction mechanism, synthesis, and physico-mechanical properties of fly ash-based geopolymers and slag-based alkali-activated binders. The literature review has been divided into following parts: [a] reactivity of precursors with variation of base materials and alkali, [b] parametric study on the performance of alkali-activated composites, [c] alkali-activated composites with supplementary calcium compounds, and [d] durability of alkali-activated composites. Based on the literature review, it is observed that there is very limited literature available on the effect of synthesis parameters on fly ash–slag blended geopolymers. The durability of fly ash slag-based geopolymers seems to have attracted less attention in the past. Finally, a brief summary has been presented, highlighting the need for future research on fly ash-based geopolymers especially in the presence of slag.

2.3.1 REACTIVITY OF PRECURSORS WITH VARIATION OF BASE MATERIALS AND ALKALI

In the 1950s, Victor Glukhovsky developed alkali-activated binder systems in Ukraine. Later on, Pavel Krivenko developed the system further. During this, the alkali-activated system was used in the construction of a tall building in Ukraine. Glukhovsky was the first one to suggest that the new cementitious system can be modeled based on the transformation of volcanic rocks to zeolite at high temperature and pressure. At first, Glukhovsky and Krivenko called the systems "soil silicate" but later on

Davidovits named them geopolymers and classified them as the three-dimensional aluminosilicates that are formed at low temperature and short time by naturally occurring aluminosilicates (Davidovits) [69]. Duxson et al. described the material as inorganic polymers. During geopolymerization, the aluminosilicate source material reacts with alkaline activator solution to form a paste which quickly transforms into a solid (Phair) [70].

The development of alternative binders based on the alkali-activation of blast furnace slag has a relatively long history. Feret first reported the use of AAS in cement in 1939, while Purdon [71] used blast furnace slag activated with sodium hydroxide to produce the binders. According to him, the polymerization process occurred in two steps. During the first step, liberation of silica, alumina, and calcium hydroxide took place and in the second step, the formation of silica and alumina hydrates along with the regeneration of the alkali solution occurred. The results led him to conclude that alkali hydroxide acted as a catalyst in leaching of some quantity of original slag mixture. Glukhovsky et al. [72] developed a new type of binder that he named "soil cement". The word soil was used because it seemed like a ground rock and the word cement because of its cementitious properties. The "soil cement" was obtained from ground alumina-silicate mixed with rich alkali industrial wastes. According to Glukhovsky et al. [72] during alkali activation, the Si-O-Si and Al-O-Si bonds were broken down when the alkalinity of the system in the form of pH increased. Then through coagulation and condensation, new stable structural units are formed.

Investigation in the field of alkali activation of aluminosilicate materials had an increase after the French organic chemist Joseph Davidovits [74] in 1978. He developed and patented binders prepared for the alkali activation of metakalolin and named it "Geopolymer". According to him, geopolymerization is a complex multi-phase exothermic process involving a series of dissolution – reorientation-solidification reactions analogous to zeolite synthesis. The geopolymerization process involves the chemical reactions under highly alkaline conditions on Si-Al minerals that results in a three-dimensional polymeric bond consisting of Si-O-Al-O bonds, represented as follows:

$$M_n\left[-\left(SiO_2\right)_z - AlO_2\right]_n . wH_2O$$

where M = the alkaline element or cation such as potassium, sodium, or calcium; the symbol – indicates the presence of a bond; n is the degree of polymerization; z is 1, 2, 3, or higher. "Geopolymer" is used to describe specifically the zeolite-like alkali-aluminosilicate materials. In this work, AAS is used to describe cementitious materials made from the activation of blast furnace slag by alkali and silicate solutions. Lee et al. [75] compared alkali activation of two different sources, namely, a calcium-rich source (slag) and an aluminosilica-rich source (fly ash) with the gel characteristics of the former being similar to a C-S-H gel and the latter material formed a polymeric structure.

John L. Provis and Jannie S.J. van Deventer [76]

In this paper, the authors used a polychromatic synchrotron beam to understand the reaction kinetics of early age geopolymerization. The study mainly focused on the effect of various synthesis parameters and manufacturing conditions such as the SiO_2/Al_2O_3 ratio, reaction temperature, and Na/ (Na+K) ratio. The study suggests that increasing the SiO_2/Al_2O_3 ratio can slow down the initial rate of geopolymerization and can also lead to a sudden pause in the reaction process if the ratio is too high. The authors also developed an equation correlating curing temperature with the specimen composition to calculate the activation energies of geopolymeric specimens.

Hua Xu and J.S.J. Van Deventer [77]

The authors used 15 different natural aluminosilicate source materials to manufacture geopolymer specimens. They were supplied by the "Geological Specimen Supplies", Turramurra, NSW, Australia. The study revealed that irrespective of the source material, increasing the concentration of the alkaline activator solution leads to a greater degree of dissolution and that sodium hydroxide was better in inducing it than potassium hydroxide. The distinct relationship between the rate of dissolution of aluminum and silica ions from the source materials was also observed. A correlation coefficient of 0.93 was calculated by the authors when comparing the rates of dissolution of the Si and Al ions.

P. De Silva, K. Sagoe-Crenstil, and V. Sirivivatnanon [78]

In this study, the reaction kinetics of metakaolin-based geopolymer specimens activated with both sodium silicate and sodium hydroxide solutions were reported. The study focused on the change in the properties of the fresh specimen and mechanical and microstructural features of the hardened specimen due to the variation of the SiO_2/Al_2O_3 ratio, curing temperature, and curing duration. It was seen that a lower SiO_2/Al_2O_3 ratio resulted in incomplete dissolution of the source material. The incomplete dissolution leads to an improper geopolymerization which resulted in lower strength. It was also seen that an increase in the aluminum content increased the rate of crystallization.

Ubolluk Rattanasak and Prinya Chindaprasirt [79]

The study reported on the rate of leaching of fly ash particles. The amount of SiO_2 and Al_2O_3 being leached was recorded. It was seen that the increase in NaOH concentration increased the rate of leaching of the Al^+ and Si^+ ions. There was also a variation in the mixing procedure where for one set of tests separate mixing of fly ash and sodium hydroxide was performed and then the sodium silicate solution was added later on and for another set, the sodium silicate and sodium hydroxide were mixed together with fly ash. It was seen that when fly ash was mixed separately with sodium hydroxide, the rate of dissolution of the fly ash particles increased. The leaching rate was also seen to depend upon the leaching time. The effect of leaching rate on the microstructure can be observed from the SEM images above. The optimum leaching time was seen to be 10 mins, and beyond that, there was no substantial increase in the rate of dissolution. High-strength geopolymers of 60–70 Mpa were reported at 10 M and 15 M NaOH.

Ailar Hajimohammadi, John L. Provis, and Jannie S.J. van Deventer [80]

The authors reported on the role of silica availability in controlling the rate of geopolymerization process. The study involved samples with a similar chemical composition but different silica availability. Sodium silicate is added to function as an external silica source. It was seen that in the samples with higher silica availability, there were more contributions from the aluminum ions to the creation of the geopolymeric gel network. This was explained by the fact that as the percentage of silica increased, the location of the nucleation sites was away from the unreacted silica particles. Spatially resolved infrared microscopy also confirmed the extent of silica dissolution. The FTIR spectra where the peak due to Si-O-Si bonds decreases with the increase of the Na_2O/SiO_2 ratio indicate that as the silica to alumina ratio increases there is an increase in the precipitation of alumina ions rather than silica ions.

Sindhunata, J.S.J. van Deventer, G.C. Lukey, and H. Xu [81]

In this study, the effect of variation of curing temperature and silicate concentration is studied. It is seen that the amount of mesopores in the specimen increases with an increase in curing temperature. The silicate ratio is varied from 0.0 to 2.0. The water to fly ash ratio was kept constant at 0.3. It is seen that with an increase in silicate concentration, there is an improvement in the pore structure of the geopolymer specimen. However, when the ratio is close to 2.0, then the degree of reaction decreases. Both sodium- and potassium hydroxide-based activator solutions were used in the manufacturing of the specimens and it was seen that the strength of the sodium-based geopolymers increased at a faster rate than that of the potassium-based one.

Sanjay Kumar and Rakesh Kumar [82]

Kumar et al. studied the effect of mechanical activation of fly ash on the various mechanical and microstructural features of the alkali-activated geopolymer specimens. It was seen that reactivity of the fly ash particles increases after mechanical activation of the particles. It was seen that as the particle size of the fly ash decreased due to vibratory milling, the external energy requirement of the geopolymer system decreased allowing the reaction to take place at room temperature. FTIR analysis revealed that mechanical activation of the fly ash particles led to the Si-O stretching bond to shift from a lower wavelength signifying a polymeric structure of a lesser order to a higher value denoting a higher degree of polymerization.

Wang S.D. and Scrivener K.L [83]

The microstructural development was studied during alkaline activation of slag pastes. Different types of slags were used in this study, and they were activated with sodium hydroxide and sodium silicate solution. Different tests such as XRD, SEM, and differential thermal analysis were conducted on the samples, and the authors came to a conclusion that alkali activation of slag took place through a mix of early age dissolution and precipitation followed by a solid-state mechanism at a later stage.

Shi and Day [84]

In this study, the effect of the dosage of activator solution on the properties of alkali-activated composites was studied. Different activator solutions in the form of sodium hydroxide, sodium silicate, and sodium carbonate-based solutions were used. The effect of the various solutions on the various stages of the alkali activation was studied. Among the above-mentioned solutions, it was seen that a higher dosage of sodium hydroxide and sodium silicate has a more pronounced impact on the period of induction and setting time of the specimens than sodium carbonate solution.

2.3.2 PARAMETRIC STUDY ON THE PERFORMANCE OF ALKALI-ACTIVATED COMPOSITES

Gum Sung Ryu, Young Bok Lee, Kyung Taek Koh, and Young Soo Chung [85]

The effect of the increase of NaOH content on the mechanical properties of geopolymers was investigated in this paper. The ratio of NaOH to sodium silicate was taken as 0:100, 25:75, 50:50, 75:25, and 100:0. Results point to the fact that the highest compressive strength of value 47 MPa was achieved at a ratio of 1:1. The SEM images also help us in understanding the complete breakdown of the fly ash particles at a higher alkali content and also the existence of partially reacted fly ash particles at low alkali concentration.

Dimitrios Panias, Ioanna P. Giannopoulou, and Theodora Perraki [86]

The effects of various synthesis parameters such as water content, alkali content, and silica content on the properties of geopolymers were studied. Analytical techniques such as X-ray diffraction and FTIR spectroscopy analysis were adopted for explaining the microstructural features and changes in reaction mechanisms. A 2.3 M concentration of soluble silicate resulted in a compressive strength of 41.3 Mpa. A compressive strength of 24.5 Mpa was recorded against a NaOH concentration of 6.6 M. Lesser amount of water content led to an increase in the compressive strength with 24.5 Mpa being recorded for a solid to liquid value of 2.05 gm/ml.

Gökhan Görhan and Gökhan Kürklü [87]

In this study, Class F fly ash from a thermal power plant in Kütahya Seyitömer (Turkey) has been used. The experimental study was designed to find a relationship between alkaline concentration and curing temperature. Thus, NaOH concentration as well as the curing temperature was varied. A multi-compound activator solution consisting of a sodium silicate solution of a fixed concentration was mixed with three separate solutions consisting of three molarities of sodium hydroxide concentrations (3 M, 6 M, and 9 M). The optimum concentration of sodium hydroxide was found to be 6 M. The bulk density values increased with the increase of temperature from 65°C to 85°C. The increase in temperature leads to improvement in the porosity of the specimens.

Peter Duxson, John L. Provis, Grant C. Lukey, Seth W. Mallicoat, Waltraud M. Kriven, and Jannie S.J. van Deventer [88]

In this paper, the authors have proposed a mechanistic model correlating the change in the Si/Al ratio with the change in mechanical strength and microstructural features. Different concentrations of alkali solution of R=2.0, 1.5, 1.0, 0.5, and 0.0 were used for the experimental procedure. The compressive strength was seen to increase with the increase in the Si/Al ratio from 1.15 to 1.90. Though the Young's modulus increases, it is only marginally from the Si/Al ratio of 1.65 to 1.90. The major change in the microstructure occurs from the Si/Al ratio of 1.4 to 1.65 with pore volume reducing drastically.

Prinya Chindaprasirt, Pre De Silva, Kwesi Sagoe-Crentsil, and Sakonwan Hanjitsuwan [89]

This study dealt with the effect of externally added silica and alumina on the properties of Class C fly ash. Nanosilica and nanoalumina are added to the system. The silica to alumina ratio had an effect on the setting time of the mix which decreases up to an optimum silica to alumina ratio. In the XRD graph, it was seen that there is presence of C-S-H phase and as the percentage of silica increases in the system, the concentration of calcium ions in the system is also seen to increase. It was seen that the geopolymerization mechanism in a high calcium fly ash-based system is different from that of a normal Class F fly ash-based system. A set of reactions were suggested by the authors for high calcium fly ash-based geopolymer reactions, which are given below.

Ravindra N. Thakur and Somnath Ghosh [90]

In this study, the effect of the synthesis parameters on the mechanical as well as microstructural features of fly ash-based geopolymer composites was reported. The Na_2O/Al_2O_3, SiO_2/Al_2O_3, and water to binder ratios were varied. It was seen that the specimen with the most alkali content recorded the highest compressive strength of 48.20 Mpa. However, in case of effect of silica content, it was seen that the optimum content of silica is 4.0 beyond which there was a decrease in strength. It was also seen that the compressive strength increased with air curing. Geopolymer mortars were also cast in this study. There was no decrease in strength relative to the paste specimen up to fly ash: sand ratio of 1:1, beyond that a decrease in strength was noted.

Behzad Nematollahi and Jay Sanjayan [91]

In this study, Class F fly ash was activated with a multi-compound activator consisting of 8 M NaOH solution and sodium silicate solution and mixed with superplasticizers (SPs). Six different types of admixtures consisting of melamine, naphthalene, and polycarboxylate were used. It was seen from the results that the effect of the SPs depended on the nature of the activator solution. In the case of NaOH-based solution, the naphthalene-based admixture had a better effect. When the fly ash was activated with a multi-compound activator, the polycarboxylate admixture was found to be more suitable as it recorded an increase in slump of 39–45% relative to the fly ash paste without SP.

Mohammed Nadeem Qureshi and Somnath Ghosh [92]

Here, the changes due to the addition of silica were studied with respect to mechanical, microstructural, and mineralogical changes. Based on the results, it was seen that there was an improvement in the compressive strength and ultrasonic pulse velocity of the specimens up to an optimum silicate ratio of 0.8. MIP was used to measure the porosity by analyzing the pore size distribution, total pore volume, etc. It was seen that when the silicate ratio was varied from 0.2 to 0.8, the compressive strength increased. The same trend was noticed for ultrasonic pulse velocity values. The microstructural characteristics were also seen to improve with increasing silicate ratio up to 0.8 with lesser voids being detected in the images.

Maochieh Chi [93]

In this study, the alkali content ($Na_2O\%$) was varied, and its effect on split tensile, compressive strength, drying shrinkage, and electrical conductivity was measured. The specimens were also subjected to sulphate and elevated temperature exposure. It was seen that when the samples were subjected to curing at 80% relative humidity at a temperature 60°C, the mechanical properties as well as its resistance against elevated temperature and sulphate exposure increased. Lime water curing was also observed to be beneficial to durability characteristics of the specimens.

Susan A. Bernal, Ruby Mejía de Gutiérrez, Alba L. Pedraza, John L. Provis, Erich D. Rodriguez, and Silvio Delvasto [94]

In this study, Colombian ground granulated blast furnace slag (GGBFS) acquired from Acerías Paz del Río has been used as the sole binder. The activator solution was composed of sodium silicate solution with a silica modulus of 2.4. The performances of the samples were compared with the increase in binder content which led to an increase in compressive strength. The trend remained the same for the OPC samples. The binder content was given at a rate of 300 kg/m³, 400 kg/m³, and 500 kg/m³. The rate of change of increase was more for the 400 kg/m³ sample than for the 500 kg/m³ sample. The curing period was also extended to 90 days for all samples. There was a marked decrease in the porosity of the AAS samples relative to the OPC samples.

Vladimir Živica [95]

In this study, three different activator solutions, namely, NaOH, Na_2CO_3, and Na_2SiO_3, have been used as activators. For all the activators, the setting time and the workability decreased when the dosage was increased in the range of 3%, 5%, and 7%, respectively. It was seen that the availability of silicate ions increased the rate of dissolution of geopolymerization process. The rate of geopolymerization was seen to be less in comparison to sodium silicate when the activator solution with sodium carbonate was used.

Susan A. Bernal, Rackel San Nicolas, Jannie S.J. van Deventer, and John L. Provis [96]

Here, a blend of sodium carbonate and sodium silicate has been used as an activator solution with granulated blast furnace slag acting as the sole source material. The specimen was cured at ambient temperature. Isothermal calorimetry results suggested that sodium silicate solution led to high dissolution rates. Carbonate ions in the solution were consumed by slag to form calcium carbonate which contributed to the early stage hardening of the specimen. The researchers suggested that the alkali carbonate salts contribute to the early age strength gain and the calcium alumina–silicate hydrate gel, hydrotalcite-type phases contribute to the ultimate strength. The authors reported the highest compressive strength of more than 60 Mpa at an age of 56 days.

Serdar Aydın and Bülent Baradan [97]

In this experimental investigation, the authors used Turkish slag as a source material to manufacture AAS composites. Physico-mechanical tests such as setting time, workability, and compressive strength were conducted. Based on the experimental results, the authors came to a conclusion that physico-mechanical characteristics of the specimens were linked to the SiO_2/Na_2O ratio. The microstructural features also depended on the SiO_2/Na_2O ratio. It was seen that increasing the alkali concentration led to an increase in compressive strength. The early age strength of the AAS mortars was lower than that of the Portland cement mortars, which were also cast for comparison. However, it was seen that the later day strength gain of the AAS mortars was greater than that of Portland cement-based mortars. Mineralogical investigation by XRD pointed the existence of C-S-H gel, though the cohesiveness of the microstructure of the gel decreases on increasing the silicate modulus beyond an optimum limit.

2.3.3 ALKALI-ACTIVATED COMPOSITES WITH A SUPPLEMENTARY CALCIUM COMPOUND

Wei-Chien Wang, Her-Yung Wang, and Ming-Hung Lo [98]

In this study, a blend of fly ash and slag has been used as a source material. The activator solution is made of a mix of sodium hydroxide and sodium silicate solution. The fly ash was replaced by slag at a percentage of 0%, 20%, 40%, and 60%. The results indicated that the workability increases with the addition of fly ash. The increase in alkaline solution also increases the workability. There was a decrease in the setting time when the amount of alkaline solution was increased to 1.5% from 0.5% and 1%. However, the greater replacement by fly ash resulted in a decrease in compressive strength. Thus keeping the fly ash content at an optimum and increasing the alkaline content, one can produce a geopolymer with high workability as well as strength. A highest compressive strength of 93.06 Mpa was obtained by the authors.

Tanakorn Phoo-ngernkham, Akihiro Maegawa, Naoki Mishima, Shigemitsu Hatanaka, and Prinya Chindaprasirt [99]

In this study, geopolymers manufactured from different source materials, for example, fly ash, slag, and a blend of fly ash and slag, were compared based on their physico-mechanical properties. A comparison was also made between properties of different specimens based on the type of activator solution used, for example, NaOH, Na_2SiO_3, and a mix of NaOH and Na_2SiO_3. The mixtures were all cured at an ambient temperature of 23°C. It was seen that the addition of slag to the mix resulted in the formation of a calcium silicate hydrate in addition to the amorphous sodium alumina–silicate hydrate gel phase. Comparatively, it was seen that the mix with a higher content of slag and activated with the help of a multi-compound solution consisting of NaOH and Na_2SiO_3 performed better.

N. Marjanović, M. Komljenović, Z. Baščarević, and V. Nikolić, R. Petrović [100]

In this study, a mix of slag and fly ash was used as a source material and activated by a sodium silicate-based activator solution. Mortars were cast using the binder: sand ratio of 1:3. It was seen that the blend proportions affected final physico-mechanical properties. Here, the specimens activated with an activator solution of comparatively high concentration solution (10% Na_2O) had a longer setting time. The curing temperature was seen to be a crucial factor for controlling the drying shrinkage of the mix. The specimens subjected to a curing temperature of 95°C reported a much lesser degree of drying shrinkage. The microstructural features, especially the gel composition, depended on the percentage of slag in the mix. It was seen that the mix with up to 50% of slag showed the clear presence of a N-A-S-H gel and when the slag was increased up to 75%, the gel composition pointed the presence of a C-N-A-S-H hybrid gel phase.

J.G. Jang, N.K. Lee and H.K. Lee [101]

In this study, superplasticized geopolymer mix using a mix of slag and fly ash was used. It was seen that the increase in the percentage of slag led to an increase in the amount of autogenous shrinkage with the specimens with 70% and 100% slag reporting the highest value. A polycarboxylate-based SP did not have any effect on the heat of hydration but its addition led to an increase in workability. The optimum content of a polycarboxylate-based SP was seen to be 2%. Based on the results, it was seen that the amount of percentage of slag had a more profound effect on the microstructural features than the dosage of SP.

2.3.4 DURABILITY OF ALKALI-ACTIVATED COMPOSITES

Susan A. Bernal, Ruby Mejía de Gutiérrez a, and John L. [102]

In this study, only GGBFS obtained from Acerías Paz del Río, Columbia was used as a source material. It was seen that with an increase in the SiO_2/Al_2O_3 (S/A) ratio, there was an increase in seventh day strengths with the compressive strength of S/A of 4.0 and 4.4 recording double the strength of the samples with a S/A ratio of 3.6. It was also seen that samples with the S/A ratio recorded a strength gain of 83% at 180 days relative to the seven-day cured sample. When the samples were subjected to the flexural strength test, the recorded values exceeded the values calculated according to ACI 318. The results of the rapid chloride penetration test showed that the values of samples did not change much from 28 days to 90 days.

H.M. Khater [103]

In this study, the source material used for alkali activation was a blend of GGBFS, silica fume (SF), air cooled slag (ACS), and cement kiln dust (CKD). The activator solution was a blend of NaOH and Na_2SiO_3. The water to binder ratio was taken as 0.3 at 100% RH. The samples were exposed to HCl and HNO_3 of concentration of 4 M. The specimens with silica fume showed the greatest level of resistivity against acid solutions. The samples with 25% ACS suffered a greater level of deterioration with surface and edge corrosion. In comparison, the only deterioration suffered by the SF samples was at the edges. The highest level of deterioration was recorded by the samples containing 25% CKD.

N.K. Lee and H.K. Lee [104]

In this study, the resistance of a fly ash–slag-based alkali-activated composite on exposure to sulphuric and chloride acid solutions was studied. It was seen that upon addition of slag, there was a formation of C-A-S-H (calcium aluminoSilicate hydrate) gel in addition to N-A-S-H (sodium aluminosilicate hydrate), and this led to an increase in the void content of the specimens. The samples S-0 and S-10 with a slag percentage of 0% and 10% recorded the worst performance against acid exposure. The chloride ion penetration potential was also seen to decrease with the addition of slag, the main reason being the resultant decrease in porosity.

2.3.5 SULPHATE RESISTANCE OF ALKALI-ACTIVATED COMPOSITES

M. Komljenovic, Z. Baščarević, N. Marjanović, and V. Nikolić[105]

In this study, the samples have been exposed to an external sulphate agent, for example, sodium sulphate. The concentration of the solution was kept at 5%. The sulphate resistance of AAS composites was compared with that of samples manufactured from ordinary Portland slag cement (CEM). The samples were exposed for a period of 90 days. It was seen that the loss in strength of the CEM specimen was much more than that of the AAS specimens. This lesser loss in strength could be attributed to the fact that there was lesser aluminum and portlandite available in the AAS specimen to participate in the sulphate reaction.

T. Bakhareva, J.G. Sanjayana, and Y.-B. Cheng [106]

In this experimental study, the AAS specimens were exposed to sodium sulphate and magnesium sulphate solutions. The performance of the specimens was compared to that of OPC-based specimens. It was observed that the strength loss in the case of AAS and OPC specimens was 17% and 25%, respectively, when exposed to sodium sulphate solution. In the case of magnesium sulphate solution, the strength loss was 37% and 23% for AAS. It was observed that the magnesium sulphate ions permeated into the gel phase causing decalcification and in the process more harm than the sodium sulphate solutions. The main products formed because of sulphate attack in both OPC and AAS concrete are gyp.

Maurice Guerrieri, Jay Sanjayan, and Frank Collins [107]

In this experimental study, the AAS specimens were exposed to elevated temperature up to 1200°C. Its performance was compared against that of the OPC, blended slag cement paste. A great percentage of mass loss of the specimens occurred in the temperature range of 100 and 200°C. This occurred because of the evaporation of the trapped water. This also resulted in high thermal shrinkage (25%) for the AASP which was higher than that of the OPC sample (12%). The shrinkage led the propagation of extensive microcracks throughout the AASP specimens. In the AASP specimens of greater size, different thermal gradients of the outer and the inner core resulted in greater damage.

Pavel Rovnaník, Patrik Bayer, and Pavla Rovnaníková (Rovnaník, Pavel, Patrik Bayer, and Pavla Rovnaníková) [108]

In this study, the AAS samples were subjected to high temperatures from 200 to 1200°C for a duration of 1 hour. The mechanical properties such as compressive and flexural strength suffer a negative change. This change occurs largely because of shrinkage of the specimen and its increased porosity. Furthermore, during the mineralogical investigation, it was seen that at 600°C, akermanite, diopside, and wollanstonite crystals are formed. This microstructural occurrence leads to an increase in the mechanical properties especially flexural strength which increases 180% with respect to the reference value.

2.3.6 SUMMARY

The detailed chemistry of alkali activation is still the subject of much discussion in the scientific literature and depends on the nature of both the solid precursor and the alkali activator selected. However, attention is required on blended systems where a mix of alumina, silica, and calcium ions is activated with alkali metal hydroxide and silicate solutions. These materials generally form an alkali cation-based amorphous gel binder.

During the course of this review, it was seen that the activator concentration and its nature affect the dissolution rate of the silicon, aluminum, and calcium [109,84], thus influencing the nature of the final reaction products. Specific activator concentration can lead to the formation of a gel structure with a high amount of cross linkage [100]. On the other hand, geopolymerization of a silica-rich material leads to the formation of a three-dimensional tetrahedral gel structure. The nature of the gel formed due to the alkali activation of a blended source material is an important area of research. The long-term behavior of the alkali-activated blended geopolymer binder is even more complicated because of the latent hydraulic nature of the slag. Durability studies regarding geopolymers, manufactured from fly ash in the presence of slag, are not extensively available. The behavior of fly ash-based geopolymers in the presence of slag under different curing conditions seemed to have received less attention in the past. Again, fly ash and blast furnace slag from different sources showed different levels of reactivity under a particular condition and affect the final properties. There is a lack of understanding with regard to fly ash based-geopolymers in the presence of slag in terms of optimal content and concentration of ingredients and their performance when exposed to acid/sulphate solutions and elevated temperature. The following chapters attempt to fulfill the abovementioned gaps especially for fly ash-based alkali-activated composites in the presence of slag.

3 Physical, Mechanical, and Microstructural Properties of Alkali-Activated Paste and Mortar

3.1 PREAMBLE

This chapter presents the details of systematic experimental investigations conducted for studying the engineering properties of blended (fly ash + blast furnace slag) alkali-activated composites (AACs) in the fresh and hardened states. The main objective of presenting these details is to have a deeper understanding of the influence of Na_2O, SiO_2, water/binder (w/b) ratio, curing temperature, curing duration, and fineness of particles on physico-mechanical properties and microstructure of fly ash-based AACs as well as their preparation and characterization. Experimentation was conducted in the Concrete laboratory of Construction Engineering as well as in the laboratories of Metallurgical and Material Engineering Department of Jadavpur University, Kolkata, India, Indian Association for Cultivation of Science, UGC DAE-CSIR, Kolkata, and Central Glass and Ceramic Research Institute (CGCRI), Kolkata, India. For characterization of fly ash-based AACs, mineralogical/microstructure/elemental analysis and pore structure studies were carried out using scanning electron microscopy (SEM)/X-ray diffraction (XRD)/energy dispersive X-ray (EDX)/Fourier transform infrared spectroscopy (FTIR)/thermogravimetric analysis (TGA), and mercury intrusion porosimetry (MIP) techniques. The laboratory tests were conducted as per relevant Indian standard codes and ASTM standards.

3.2 SOURCE MATERIALS

3.2.1 FLY ASH

Typical Class F fly ash is the main base material used for the manufacturing of alkali-activated composite specimens throughout the research. The lignite coal-origin dry fly ash was obtained from the Kolaghat Thermal Power Plant, located near Kolkata, India. The fly ash was kept in a plastic container with a lid to avoid any contact with moisture, etc. The chemical oxide composition of the fly ash, obtained using X-ray fluorescence (XRF) analysis, particle size distribution, and SEM images, is given in Table 3.1, Figure 3.1, and Figure 3.2, respectively.

TABLE 3.1

Chemical Composition of Fly Ash by XRF

Chemical Composition	SiO_2	CaO	Al_2O_3	MgO	Fe_2O_3	S	Na_2O	K_2O	LOI
Mass (%)	65.81	2.45	22.17	0.35	3.23	0.47	0.65	0.85	0.81

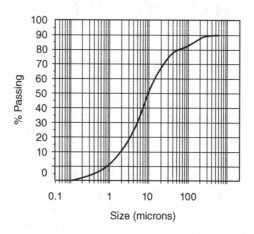

FIGURE 3.1 Particle size distribution of fly ash.

FIGURE 3.2 SEM micrograph of fly ash.

Thus, as per ASTM standard C6128-03, the fly ash can be described as Class F fly ash (or as per IS 3812(Part-I)-2003 specifications, siliceous pulverized fuel ash). The color of fly ash was grayish black.

3.2.2 GROUND GRANULATED BLAST FURNACE SLAG (GGBS)

GGBS used in this investigation was obtained in granular form (4–5 mm) from Tata Metaliks Ltd., Kharagpur, India. GGBS obtained was about courser having an irregular popcorn-like shape with observable pores as shown in Figure 3.3. Slag was ground in a mechanical grinder to particle size less than 45 μm. Then, it was stored in a sealed plastic container, to ensure the consistency in the material. The GGBS was kept in the plastic container with a lid to avoid any contact with moisture, etc. The chemical composition of the GGBS was determined by XRF and is given in Table 3.2. Its specific gravity was 2.90 with particle size less than 45 μm.

Figure 3.4 shows the particle size distribution of GGBS obtained by the laser diffraction technique. The measurement instrument used was a Microtrac make S3500 particle size analyzer. About 100% particles were smaller than 45 μm. The SEM image and EDX of GGBS are shown in Figures 3.5 and 3.6.

FIGURE 3.3   Photograph of raw GGBS.

TABLE 3.2
Chemical composition of blast furnace slag obtained by XRF

Chemical Composition	SiO$_2$	CaO	Al$_2$O$_3$	MgO	Fe$_2$O$_3$	S	Na$_2$O	K$_2$O	LOI
Mass (%)	37.25	42.17	10.24	3.82	1.1	2.1	0.19	0.66	0.81

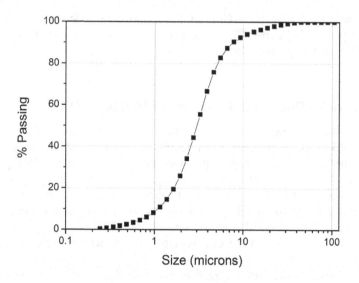

FIGURE 3.4   Particle size distribution of GGBS obtained by laser diffraction analysis.

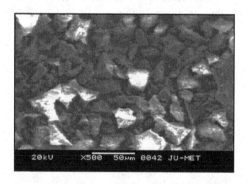

FIGURE 3.5   SEM image of GGBS.

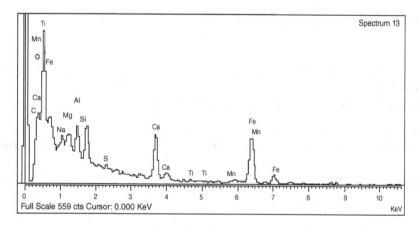

FIGURE 3.6   EDX spectrum of GGBS.

3.2.3 ACTIVATOR SOLUTION

The activator solution was the combination of Na_2SiO_3 solution and NaOH pellets. Laboratory-grade NaOH pellets (97% purity with $Na_2O = 77.5\%$ and 22.50% water) were acquired from Merck India Ltd. and Na_2SiO_3 solution ($Na_2O = 8\%$, $SiO_2 = 26.50\%$ and 65.50% H_2O) with silicate modulus ~ 3.3 and a bulk density of 1410 kg/m^3 was acquired from Loba Chemie Ltd. India.

3.3 MANUFACTURING PROCESS OF BLENDED AACS

Blended AAC paste is prepared by mixing only fly ash or a blend of fly ash and slag with alkaline activator while an AAC mortar are prepared by adding fine aggregates, that is, sand into the mix. Both paste and mortar show a cohesive nature in the fresh state. The following constituents were used in the manufacture of AACs for the present research work.

Alkali content ($\%Na_2O$) = 4%–12% by weight of binder material
Silica content (% SiO_2) = 4%–14% by weight of binder material
w/b material ratio = 0.35–0.47
Sand to binder material ratio = 1 by weight
Fineness of GGBS = <45 to < 75 μm
Fineness of Fly ash = <45 μm

The following manufacturing process was adopted for preparing the blended AACs.

A. Preparing the activator solution, alkaline in nature by mixing NaOH pellets, sodium silicate solution, and water according to mix proportion, to make alkaline activator, at least one day prior to its use.
B. Dry mixing of fly ash and GGBS followed by addition of activator solution into a Hobart mixer for about five minutes to obtain homogeneous paste.
C. For preparing mortar, dry mixing of fly ash, GGBS and sand followed by addition of activator solution into a Hobart mixer for about five minutes to obtain homogeneous mix.
D. Measuring the flow of the mix (paste or mortar) using mini slump cone apparatus.
E. Then the paste was transferred onto 50 mm × 50 mm × 50 mm cube mold and cast in two layers with each layer subjected to manual compaction.
F. Then, the molds were transferred to the vibrating table to achieve adequate compaction.
G. A rest period of 24 h is given to water cured specimens were taken out from the molds and kept at fully submerged condition in water at ambient temperature until the date of testing, while a rest period of 4 h is given for heat-cured specimens prior to placing them in an oven for thermal curing for 24 h at 40°C–85°C. After de-molding, the specimens were left to air drying at room temperature until the day of testing.

3.4 TESTING PROCEDURES AND CHARACTERIZATION OF BLENDED AACS

The basic goal for testing of AACs is to investigate the engineering properties and to characterize the mineralogy and microstructure of the specimens manufactured with different fly/slag ratios, and different Na_2O and SiO_2 contents under different test conditions. The laboratory tests were conducted as per relevant Indian standard codes and ASTM guidelines. For mineralogical and microstructural characterization, elemental analysis, and pore characteristic measurement of AACs were carried out using XRD, SEM, EDX, differential thermal analysis (DTA)/TGA, and MIP. The details of various test procedures used in the present research are described in the following section.

3.4.1 WORKABILITY/FLOW

Workability was measured using a mini flow table. The test was carried out as per ASTM C 1437-07 but with a slight modification; the table was raised and dropped 15 times in about 15 s (Figure 3.7). The apparatus was checked as per ASTM C 230/C 230M. The workability was assessed by taking the average value of the diameter of the flowing paste in two directions perpendicular to each other. Depending on the flow diameter, workability of the AAC mix was classified as stiff, moderate, and high.

3.4.2 SETTING TIME MEASUREMENT

The setting time was measured using a Vicat apparatus (Figure 3.8) as per IS 5513-1996, and the initial and final setting time procedures were adopted as per ASTM C191-08 and IS 4031 (Part 5)-1988. A cylindrical container of 75 mm diameter was filled with 50 mm of AAC paste. Values of penetration distances were measured every 10 mins. For convenience, initial setting time is regarded as the time elapsed between the moments that the extra water is added to the binder, to the time that the paste starts losing its plasticity. The initial setting time is the time at which the needle of 1 mm dia. penetrates the paste in a cylindrical container to a depth equal to 35 ± 0.5 mm from the top. The final setting time is the time elapsed between the moment the extra water is added to the AAC binder and the time when the paste has completely lost its plasticity and has attained sufficient hardness to resist certain definite pressure. In other words, final setting was the time required to reach a penetration of zero mm.

The workability of the fresh AAC paste specimen was studied using a mini flow table test as per ASTM C1437-07, with one modification: the table was raised and

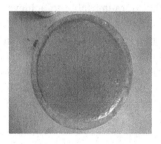

FIGURE 3.7   Measurement of workability using mini flow table test apparatus.

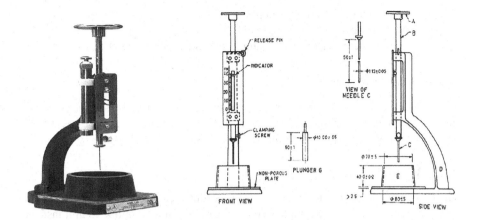

FIGURE 3.8 Measurement of Setting time using Vicat apparatus. [IS 5513-1996]

dropped 15 times in about 15 s. The reason for this modification is that some mix tends to spread more than the diameter of the table (250 mm), and the purpose of the test would be lost. The workability of the mix was determined by measuring the diameter of paste flow on a flow table in two perpendicular directions after 15 drops in 15 s, and the average value is considered as a flow diameter. The percentage increase in the flow diameter with respect to the initial diameter of mold is considered as flow value. The static diameter is the diameter of paste when mold is lifted up and paste spreads under its own weight. The dynamic diameter is the diameter of paste after 15 drops applied during the flow test. All the tests were repeated 3–5 times, and the average flow was considered.

3.4.3 Bulk Density and Apparent Porosity

The bulk density and apparent porosity were calculated for specimens cured for 28 days according to the Archimedes principle using the process mentioned below using water as the submerging medium.

The following procedure was followed for the test:

A. The weight of dry specimens was measured as 'D' after the curing period is over.
B. Specimens are then soaked in water for 24 h at room temperature.
C. The specimens were removed from water and then suspended by a thin wire inside water, and their weight was recorded as 'S'.
D. The specimens are then wiped dry and their weight was measured in the saturated surface dry condition as 'W'.

The bulk density and apparent porosity of the specimens are then determined using the relationships given below.

$$\text{Bulk density}\left(kg\,/\,m^{3}\right)=\left[D\,/\left(W-S\right)\right]\times1000$$

$$\text{Apparent Porosity}\left(\%\right)=\left[D\,/\left(W-S\right)\right]\times1000$$

Where, D = Weight of oven dried sample in kg.

  W = Saturated weight of the sample in kg.

  S = Weight of the water-saturated sample suspended in water in kg.

3.4.4 Water Absorption

Water absorption is measured by drying a specimen to a constant mass and immersing in water and measuring the increase in mass as a percentage of dry mass. Water absorption of specimens was determined as per ASTM C-642 (Figure 3.9).

3.4.5 Water Sorptivity

The sorptivity test was carried out as per the procedure described in ASTM C 1585-04 with a modification that the sample size used was 50 mm × 50 mm × 50 mm instead of 4-in. (100-mm) diameter, 2-in. (50-mm) long cylinders. The test method was used to determine the rate of absorption (sorptivity) of water by AAC hardened paste specimens by measuring the increase in mass of specimen resulting from absorption of water as a function of time when only one surface of the specimen is exposed to water, and water ingress of unsaturated specimen was dominated by capillary suction during initial contact with water. The test procedure has been carried out at an average room temperature of 28°C with tap water conditioned to the same temperature. The mass of the specimens was measured near to 0.01 g and recorded as the initial mass of test specimen. The pan having the arrangement as shown in Figure 3.10 was used in this study. The water level of 1–3 mm above the bottom of the specimens was maintained for the duration of the tests. The specimens are then weighed, and the absorbing surfaces are exposed to water by immersion in a pan containing water. The time device was started immediately after placing the test specimens in the pan. The mass of the specimens was recorded at 60 minutes ± 2 s and the second point at 5 minutes ± 10 s, and subsequent measurements were made at ± 2 minutes of 10 minutes, 20 minutes, 30 minutes, and 60 minutes. The actual time was recorded to within ± 10 s. The measurements were continued every hour ± 5minutes up to 6 hours from the first contact of the specimen with water, and time was recorded within ± 1 minute. After the initial 6 hours, measurements were taken once a day up to 3 days

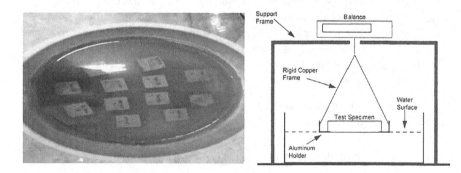

FIGURE 3.9 Measurement of water absorption, bulk density, and apparent porosity.

FIGURE 3.10 Setup for the water sorptivity test.

followed by 3 measurements 24 hours apart during days 4–7 and final measurements 24 hours after measurement at 7 days. At increasing time intervals, the specimens were removed from exposure to water, the surfaces were blotted to remove excess surface water, and the specimen was re-weighed within 15 s of removal from the pan.

The absorption 'I' is the change in mass divided by the product of the cross-sectional area of the test specimen and the density of water. The temperature dependency of the density of water is neglected in this test. The cumulative water absorption (per unit surface area) increases as the square root of the elapsed time 't'.

$$I = S\sqrt{t}$$

The test results data points were fitted into the equation

$$I = S\sqrt{t} + C$$

Where

I = Increase in mass per unit area (gm/mm²)
t = time in minutes at which the mass is determined
S = Sorptivity in gm/mm²/mm$^{0.5}$
C = constant

The initial rate of water absorption (mm/s$^{0.5}$) is defined as the slope of the line that is best fit to 'I' plotted against the square root of time (s$^{0.5}$). Regression analysis was carried out to obtain the slope of line using the least-squares method for all points from 1 minute to 6 hours with a correlation coefficient of at least 0.98. The secondary rate of water absorption (mm/s$^{0.5}$) is the slope of the line that is the best fit to 'I' plotted against the square root of time (s$^{0.5}$) using all the points from 1–7 day using least-squares linear regression to determine the slope with a correlation coefficient of at least 0.98.

3.4.6 Ultrasonic Pulse Velocity (UPV)

The UPV method was carried out as per IS: 13311 (Part 1)-1992 using a commercially available PUNDIT system as shown in Figure 3.11.

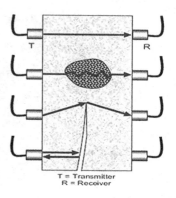

T = Transmitter
R = Receiver

FIGURE 3.11   Set up for the measurement of UPV.

FIGURE 3.12   Digital compression testing machine.

3.4.7 COMPRESSIVE STRENGTH

The AAC specimens (50mm × 50mm × 50mm) were tested for compressive strength using a digital compressive testing machine with a loading rate of 20 MPa/min. The compressive strength tests were conducted at the age of 3, 7, and 28 days. Three–six specimens of each series at each age were crushed in a digital compression testing machine in accordance with ASTM C-109-02, and the average value is reported as the compressive strength. A typical test setup for compressive strength is shown in Figure 3.12.

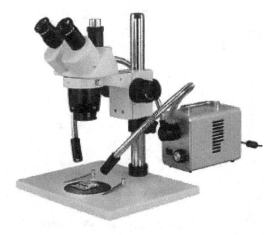

FIGURE 3.13   Optical microscope.

3.4.8 Physical Changes and Optical Microscopy

AAC specimens are examined for any noticeable changes in their physical form and for any other deposits on the surfaces when exposed to elevated temperature and acid and sulphate attacks. An optical microscope as shown in Figure 3.13 was used to observe the surface texture of the specimens at regular intervals during exposure to various conditions. Photographs of the surface texture were taken with a digital camera. Similar observations on unexposed specimen's surfaces were made for comparison purposes.

3.4.9 XRD Analysis

A Rigaku Ultima III X-ray diffractometer as shown in Figure 3.14 was used to perform XRD, a form of non-destructive analysis of materials, typically in powder or thin film form. XRD has a wide variety of applications in material characterization, such as determination of the crystal structure and orientation, phase identification, and detection of trace compounds in a material. When a crystalline mineral is exposed to X-rays of a particular wavelength, the layers of atoms diffract the rays and produce a pattern of peaks, which is a characteristic of the mineral. The horizontal scale (diffraction angle) of a typical XRD pattern gives the crystal lattice spacing, and the vertical scale (peak height) measures the intensity of the diffracted ray. In the present research, XRD analysis was performed using a Rigaku Ultima III machine (Rigaku, Japan) with Cu-Kα radiation with the following conditions: tube voltage and current – 40 kV and 30 mA. The XRD patterns were obtained in steps of 0.50 (2theta) at a rate of 1° (2 theta) per minute, sweep from 10° to 70° (2 theta), according to the powder diffraction method. XRD analysis of a few specimens was also conducted using an X'pert Pro MPD diffractometer (PANalytical) operating at 40 kV and 30 mA using Ni-filtered Cu-Kα radiation with a step size of 0.050 (2theta), a step time of 75 s, and sweep from 5° to 95° (2 theta). The results were compared with the International Centre for Diffraction Data (ICDD) database.

FIGURE 3.14   Rigaku Ultima III X-ray Diffractometer.

3.4.10 SEM AND EDX ANALYSIS

SEM and EDX analyses were carried out to make the qualitative assessment of the microstructure and elemental analysis of AACs. SEM specimens were dried and then platinum coated (8 nm) using a JEOL JFC-1600 auto fine coater (Figure 3.15) prior to examination using a JEOL JSM-6360 scanning electron microscope equipped with an Oxford Inca EDX analyzer as shown in Figure 3.16.

FIGURE 3.15   JEOL JFC-1600 autofine coater.

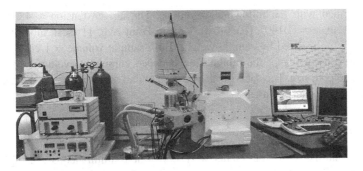

FIGURE 3.16   JEOL JSM 6360 microanalysis system equipped with an Inca Oxford EDX analyzer used for SEM/EDX analysis.

3.4.11 MIP Test

MIP is based on the fact that a non-wetting liquid (one having a contact angle greater than 90°) will only intrude capillaries under pressure.

The relationship between the pressure and capillary diameter is described by Washburn as:

$$P = \frac{-4\gamma \cos\theta}{d}$$

where P = Pressure
   γ = Surface tension of the liquid
   θ = Contact angle of the liquid
   d = Diameter of the capillary pore

Mercury must be forced to use pressure into the pores of a material. The pore size distribution is determined from the volume intruded at each pressure increment. Total porosity is determined from the total volume of mercury intruded into the test sample. The MIP technique is widely used because of its ease and simplicity. In the present investigation, a Quantachrome Poremaster 60 was used as shown in Figure 3.17. The test

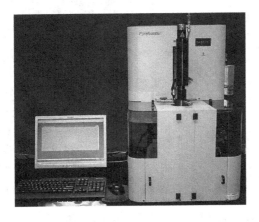

FIGURE 3.17   Quantachrome Poremaster 60 mercury intrusion porosimeter.

was performed with the following conditions: Hg surface tension = 480 erg/cm^2; Hg contact angle = (I)140°, (E) 140° ; Moving Point Avg. 11 (scan mode); mercury volume was normalized by sample weight. Bulk sample volume = 1 cc. The results obtained are used to plot the relationship between cumulative volumes of mercury intruded and pore size. The data can also be used to find the average pore diameter, total porosity, and pore surface area.

3.4.12 TGA AND DTA

The DTA and TGA measurements were performed simultaneously using a Pyris Diamond TG/DTA, Perkin-Elmer instrument as shown in Figure 3.18. Powdered specimens were used in the tests to ensure uniform heating of the samples during transient heating. Experiments were performed with the conditions: temperatures between 28°C and 1000°C, constant heating rate of 10°C per minute, and the nitrogen purge rate of 150 ml per minute. Platinum crucible was used with alpha alumina powder as the reference to accurately measure the loss of mass while the specimens were gradually exposed to elevated temperatures.

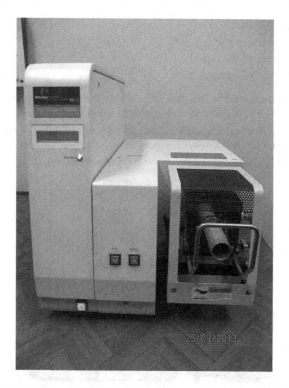

FIGURE 3.18   Pyris Diamond TG/DTA, Perkin-Elmer instrument.

FIGURE 3.19   Shimadzu IR Prestige 21 equipment.

3.4.13 FTIR

In infrared spectroscopy, IR radiation is passed through a sample. Some of the infrared radiation is absorbed by the sample and some of it is passed through (transmitted). The resulting spectrum represents the molecular absorption and transmission, creating a molecular fingerprint of the sample. FTIR spectra of blast furnace slag and hardened specimens were recorded using Shimadzu IR Prestige 21 equipment (Figure 3.19) using potassium bromide pellets (Sample: KBr = 1:50). The spectra were recorded in the range of 400–4500 cm^{-1} with 2 cm^{-1} resolution, with 40 scans recorded each time.

3.5 EXPERIMENTAL INVESTIGATION OF BLENDED AACS

3.5.1 Preamble

The objective of the research work was to investigate the effect of synthesis parameters on engineering properties and durability of blended AACs. The experimental work consists of testing various AAC paste and mortar specimens prepared by varying slag, alkali content [(% Na_2O], silica content (%SiO_2) and w/b ratio at ambient and elevated temperature and also exposed to different solutions of acid and sulphate. Compressive strength at ambient and elevated temperatures as well as after exposure to aggressive chemical environment had been determined to study the performance of AACs. In addition, simple tests such as measurement of bulk density, water absorption, water sorptivity, and UPV are conducted. The mineralogical and microstructural characterization was done using XRD and SEM/EDX, while pore characteristics of AACs have been studied using MIP.

3.5.2 Scope of Work

Fly ash and GGBS had been used as source materials for the manufacturing of AACs. The AAC mix was prepared by activating dry GGBS with a mixed solution of sodium hydroxide and sodium silicate. For preparation of AAC mortar, river sand from a local source was used as the fine aggregate. The main objective of the experimental work was as follows:

1. Preparation and characterization of AACs using fly ash and GGBS.
2. Study of effect of GGBS on engineering properties, compressive strength, and microstructure of the AACs.
3. Study of the effect of alkali concentration, silica concentration, and w/b ratio on the engineering properties, compressive strength and microstructure of the AACs.
4. Study of the effect of fineness, curing temperature, and curing duration on engineering properties, compressive strength, and microstructure of the AACs.

Based on the above objective, the experimental work was divided into two parts as follows:

1. Study of the effect of GGBS on engineering properties, compressive strength, and microstructure of the AACs.
2. Study of the effect of synthesis parameters on engineering properties, compressive strength, and microstructure of the AACs.

3.5.2.1 Study on the Effect of Addition of GGBS on Blended AACs

The aim of this part of experimental investigation was to study the effect of addition of slag on physico-mechanical characteristics and the microstructure of AACs. The AAC paste/mortar specimens were prepared by mixing dry fly ash and GGBS with an activating solution in combination of sodium hydroxide and sodium silicate and water with or without sand. The proportion of AAC mix was varied systematically by changing the quantity and proportion of GGBS with respect to the total quantity of the source material. The synthesis parameters such as alkali content (% Na_2O), silica content (% SiO_2), and w/b ratio were kept constant at 8%, 10%, and 0.38%, respectively. The compressive strength, bulk density, apparent porosity, water absorption, and water sorptivity (primary and secondary absorption) have been investigated for AAC specimens obtained at the age of 7 and 28 days and presented with respect to the percentage of slag. The obtained results have been supported and interpreted with the mineralogical and microstructural characterization performed by use of XRD, FTIR, and SEM/EDX.

3.5.2.2 Study on the Effect of Synthesis Parameters on Blended AACs

The aim of this part of experimental investigation was to study the effect of synthesis parameters on compressive strength and the microstructure of AACs. The AAC paste/mortar specimens were prepared by mixing dry fly ash and GGBS with an activating solution in combination of sodium hydroxide and sodium silicate and water. The proportion of AAC mix was varied systematically by changing the quantity and proportion of sodium hydroxide and sodium silicate and extra water in activating solution. The effects of synthesis parameters such as alkali content (% Na_2O), silica content (% SiO_2) and w/b ratio and curing conditions such as water curing and heat curing were studied. The compressive strength, bulk density, apparent porosity, water absorption, and water sorptivity (primary and secondary absorption) have been investigated for AAC specimens obtained at the age of 7 and 28 days and presented with respect to the percentage of slag. The obtained results have been supported and interpreted with the mineralogical and microstructural characterization performed by use of XRD, FTIR, and SEM/EDX.

3.5.2.3 Study on the Efect of Addition of GGBS and Synthesis Parameters on Blended AACs in the Fresh State

This part of experimental program was undertaken to study the engineering properties of AACs in the fresh state. The workability and setting time haves been investigated. Four series of experiments were conducted (slag series, alkali series, silica series, and w/b Series). In slag series, the effect of slag content on engineering properties of fresh AACs was studied. In alkali series, the effect of alkali content on engineering properties of AACs was studied. AAC specimens were prepared by varying the % (Na_2O) from 4% to 12% by keeping constant %SiO_2 equal to 8% and slag content fixed at 30% . In the third series (silica series), the effect of SiO_2 was studied on specimens prepared by varying the SiO_2 content from 4% to 14% by weight of source material by keeping % (Na_2O) constant and slag content fixed at 30% . In the fourth series (w/b series), the effect of the w/b ratio was studied on specimens prepared by varying the w/b content from 0.35 to 0.47 by the weight of the source material by keeping % (Na_2O) constant, and SiO_2 and slag content fixed at 30% . The workability and setting time of fresh AAC mix of slag series, alkali series, silica series, and w/b Series were determined as per ASTM C1437-07and ASTM-C191 standards, respectively. Depending on the flow diameter of AAC mix, the workability of AAC mix was classified as stiff, moderate, and high. The loss of flow with time was measured. The setting time of AAC mix was measured using Vicat apparatus (Figure 3.8).

3.5.3 STUDY ON VARIATION OF SLAG AND SYNTHESIS PARAMETERS ON BLENDED AACS

3.5.3.1 Mix Proportions

The following series of AAC specimens were prepared by changing the amount of slag, fly ash, sodium hydroxide (NaOH), sodium silicate solution (Na_2SiO_3), and w/b ratio in the activating solution for studying the effect of synthesis parameters on development of compressive strength and microstructure.

Test Series-1: In this series, the effect of slag content on the AAC mix was studied. The slag content was varied from 0% to 60% .The alkali and SiO_2 contents in the mix have been kept constant at 8% and 8%. Depending on the quantity of water available in the NaOH pellets and Na_2SiO_3 solution, extra water has been added into the

TABLE 3.3

Details of AAC mix for slag series

	AAC Mix Details					AAC Mix Proportion		
Mix ID	% Na$_2$O in Activator*	% SiO$_2$ in Activator*	w/b ratio*	Fly Ash (g)	GGBS (g)	Na$_2$SiO$_3$ (g)	Extra Water Added (g)	NaOH (g)
100-0	8.0	8.0	0.38	1000	0	301.89	165.55	74.29
90-10	8.0	8.0	0.38	900	100	301.89	165.55	74.29
85-15	8.0	8.0	0.38	850	150	301.89	165.55	74.29
70-30	8.0	8.0	0.38	700	300	301.89	165.55	74.29
50-50	8.0	8.0	0.38	500	500	301.89	165.55	74.29
40-60	8.0	8.0	0.38	400	600	301.89	165.55	74.29

* indicates by weight of source material (Fly ash + Slag)
* Specimen calculation available in Appendix

TABLE 3.4

Details of AAC mix for alkali series

	AAC Mix Details					AAC Mix Proportion		
Mix ID	% Na$_2$O in Activator*	% SiO$_2$ in Activator*	w/b Ratio*	Fly Ash (g)	GGBS (g)	Na$_2$SiO$_3$ (g)	Extra Water Added (g)	NaOH (gm)
AAC-4%	4%	8.00%	0.38	700	300	301.89	177.52	21.08
AAC-6%	6.0%	8.00%	0.38	700	300	301.89	171.53	47.69
AAC-8%	8.0%	8.00%	0.38	700	300	301.89	165.55	74.29
AAC-10%	10.0%	8.00%	0.38	700	300	301.89	159.56	100.90
AAC-12%	12.0%	8.00%	0.38	700	300	301.89	153.58	127.50

* Indicates by weight of source material (Fly ash + Slag).
* Specimen calculation available in Appendix.

activator solution to make up the required water to fly ash–slag ratio. The mix proportion and chemical composition of the AAC mix of this test series are shown in Table 3.3.

Test Series-2: In this series, the effect of alkali content (Na$_2$O) of the AAC mix was studied. The alkali content was varied from 4% to 12% by varying the quantity of NaOH pellets in activator solution. The SiO$_2$ content in the mix has been kept constant at 8%. Depending on the quantity of water available in the NaOH pellets and Na$_2$SiO$_3$ solution, extra water has been added in the activator solution to make up the required water to slag ratio. The mix proportion and chemical composition of the AAC mix of this test series are shown in Table 3.4.

Test Series-3: In this series, the effect of silica content (SiO$_2$) of the AAC mix was studied. The silica content was varied from 4% to 14% by changing the quantity of

TABLE 3.5
Details of AAC Mix for Silica Series

	AAC Mix Details					AAC Mix Proportion		
Mix ID	% Na$_2$O in Activator*	% SiO$_2$ in Activator*	w/b Ratio*	Fly Ash (g)	GGBS (g)	Na$_2$SiO$_3$ (g)	Extra Water Added (g)	NaOH (g)
AAC-4%	8.0%	4.00%	0.38	700.00	300.00	150.94	260.80	90.36
AAC-6%	8.0%	6.00%	0.38	700.00	300.00	226.42	213.18	82.32
AAC-8%	8.0%	8.00%	0.38	700.00	300.00	301.89	165.55	74.29
AAC-10%	8.0%	10.00%	0.38	700.00	300.00	377.36	117.92	66.26
AAC-12%	8.0%	12.00%	0.38	700.00	300.00	452.83	70.29	58.23
AAC-14%	8.0%	14.00%	0.38	700.00	300.00	528.30	22.67	50.20

* indicates by weight of source material (Fly ash + Slag).
* Specimen calculation available in Appendix.

TABLE 3.6
Details of AAC Mix for w/b Series

	AAC Mix Details					AAC Mix Proportion		
Mix ID	% Na$_2$O in Activator*	% SiO$_2$ in Activator*	w/b Ratio*	Fly Ash (g)	GGBS (g)	Na$_2$SiO$_3$ (g)	Extra Water Added (g)	NaOH (g)
AAC-4%	8%	8%	0.35	700	300	377.36	87.92	66.26
AAC-6%	8%	8%	0.38	700	300	377.36	117.92	66.26
AAC-8%	8%	8%	0.41	700	300	377.36	147.92	66.26
AAC-10%	8%	8%	0.44	700	300	377.36	177.92	66.26
AAC-12%	8%	8%	0.47	700	300	377.36	207.92	66.26

* Indicates by the weight of the source material (fly ash + slag).
* Specimen calculation available in Appendix.

sodium silicate in activator solution. The NaOH content in the mix has been kept constant at 8%. Depending on the quantity of water available in the NaOH pellets and Na$_2$SiO$_3$ solution, extra water has been added into the activator solution to make up for the required water to source material ratio. The mix proportion and chemical composition of the AAC mix of this test series are shown in Table 3.5.

Test Series- 4: In this series, the effect of water content expressed as water to AAC binder (AAC binder is the sum of the total mass of fly ash/slag + mass of solids in activating solution) was studied. For this test series, the w/b ratio was varied from 0.35 to 0.47 by changing the quantity of water in the mix. The alkali content and silica content were kept equal to 8%. The mix proportion and chemical composition of the AAC mix of this test series are shown in Table 3.6.

Test Series- 5: In this series, the effect of fineness of slag on the engineering properties and microstructural/mineralogical aspects of AACs was studied. The Na_2O, SiO_2, and w/bratio in the mix has been kept constant at 8%, 10%, and 0.38%, respectively. The mix proportion and mineralogical composition of the AAC of this test series are furnished in Table 3.7.

Test Series- 6: Details of AAC mix for curing series. In this series, the effect of thermal curing temperature and duration, on the engineering properties and micro-structural/mineralogical aspects of AAC composites, was studied. Temperature of oven curing was kept at 60°C and 85°C. The Na_2O, SiO_2, and w/b content in the mix was kept constant at 8%, 10%, and 0.38%, respectively. The duration of thermal curing was kept constant at 48 h. The thermal curing duration was varied from 24 to 72 h at an oven temperature of 60°C and 85°C. The Na_2O, SiO_2, and w/b content in the mix was kept constant at 8%, 10%, and 0.38%, respectively. Again, the effect of water curing on the engineering properties and microstructural/mineralogical aspects of AACs was studied. The water curing was carried out for 7 and 28 days. The Na_2O, SiO_2, and w/b content in the mix was kept constant at 8%, 10%, and 0.38%, respectively. Slag content varied from 10% to 60%. The mix proportion and mineralogical composition of the AAC composites of this test series are furnished in Tables 3.8–3.13.

Test Series-7: In this series, the effect of slag content, alkali content, silica content, and water content on sorptivity was studied. The mix proportion and chemical composition of the AAC mix of this test series are shown in Tables 3.14–3.17.

Series-1: Slag content was varied from 0% to 30% keeping Na_2O, SiO_2, and w/b constant at 8%, 10%, and 0.38%, respectively. Refer Table 3.14.

Series-2: Sodium hydroxide and sodium silicate solution were used as an activator. The alkali content (% Na_2O) was varied from 4% to 10% by keeping the SiO_2 content constant at 10% and maintaining w/b ratio equal to 0.38 as given in Table 3.15.

Series-3: Sodium hydroxide and sodium silicate solution were used as an activator. The silica content (%SiO_2) was varied from 4% to 14% by keeping the Na_2O content constant at 8% and maintaining w/b ratio equal to 0.38 as given in Table 3.16.

Series-4: The w/b ratio was varied from 0.35 to 0.47 by keeping the Na_2O and SiO_2 content constant at 10% as given in Table 3.17.

3.5.3.2 Tests Conducted

Workability, setting time, bulk density, apparent porosity, compressive strength, water absorption, water sorptivity, UPV, XRD analysis, SEM, EDX analysis, and FTIR.

TABLE 3.7
Details of AAC Mix for Fineness Series

							AAC Mix Details			
Mix ID	% Na$_2$O in Activator (a)*	% SiO$_2$ in Activator (b)*	w/b Ratio (c)*	Fly Ash (g)	Slag (g)	Fineness of Slag	Na$_2$SiO$_3$ (g)	Sodium Hydroxide (g)	Extra Water Added (g)
AAC-4%	8	10	0.38	700	300	<150micron	377.36	118.37	66.20
AAC-6%	8	10	0.38	700	300	75 micron–150 micron	377.36	118.37	66.20
AAC-8%	8	10	0.38	700	300	45 micron–75 micron	377.36	118.37	66.20

Notes: (a)*, (b)*, and (c)* by weight of source material (fly ash + slag).
* Specimen calculation available in Appendix I.

TABLE 3.8

Details of AAC Mix for Curing Temperature and Duration Series (Fly Ash/Slag Ratio Constant at 90/10)

Mix ID	% Na$_2$O	% SiO$_2$	w/b Ratio	Curing Temp	Curing Duration	Fly Ash (g)	Slag (g)	Na$_2$SiO$_3$ (g)	Extra Water Added (g)	Sodium Hydroxide (g)
					AAC Mix Details					
AAC-60/24	8	8	0.38	60°C	24 h	900	100	301.89	165.55	74.29
AAC-85/24	8	8	0.38	85°C	24 h	900	100	301.89	165.55	74.29
AAC-60/48	8	8	0.38	60°C	48 h	900	100	301.89	165.55	74.29
AAC-85/48	8	8	0.38	85°C	48 h	900	100	301.89	165.55	74.29
AAC-60/72	8	8	0.38	60°C	72 h	900	100	301.89	165.55	74.29
AAC-85/72	8	8	0.38	85°C	72 h	900	100	301.89	165.55	74.29

TABLE 3.9

Details of AAC Mix for Curing Temperature and Duration Series (Fly Ash/Slag Ratio Constant at 85/15)

AAC Mix Details

Mix ID	% Na_2O	% SiO_2	w/b Ratio	Curing Temp	Curing Duration	Fly Ash (g)	Slag (g)	Na_2SiO_3 (gm)	Extra Water Added (gm)	Sodium Hydroxide (gm)
AAC-60/24	8	8	0.38	60°C	24 h	850	150	301.89	165.55	74.29
AAC-85/24	8	8	0.38	85°C	24 h	850	150	301.89	165.55	74.29
AAC-60/48	8	8	0.38	60°C	48 h	850	150	301.89	165.55	74.29
AAC-85/48	8	8	0.38	85°C	48 h	850	150	301.89	165.55	74.29
AAC-60/72	8	8	0.38	60°C	72 h	850	150	301.89	165.55	74.29
AAC-85/72	8	8	0.38	85°C	72 h	850	150	301.89	165.55	74.29

TABLE 3.10
Details of AAC Mix for Curing Temperature and Duration Series (Fly Ash/Slag Ratio Constant at 70/30)

Mix ID	% Na_2O	% SiO_2	w/b Ratio	Curing Temp	Curing Duration	Fly Ash (g)	Slag (g)	Na_2SiO_3 (g)	Extra Water Added (g)	Sodium Hydroxide (g)
							AAC Mix Details			
AAC-60/24	8	8	0.38	60°C	24 h	700	300	301.89	165.55	74.29
AAC-85/24	8	8	0.38	85°C	24 h	700	300	301.89	165.55	74.29
AAC-60/48	8	8	0.38	60°C	48 h	700	300	301.89	165.55	74.29
AAC-85/48	8	8	0.38	85°C	48 h	700	300	301.89	165.55	74.29
AAC-60/72	8	8	0.38	60°C	72 h	700	300	301.89	165.55	74.29
AAC-85/72	8	8	0.38	85°C	72 h	700	300	301.89	165.55	74.29

TABLE 3.11

Details of AAC Mix for Curing Temperature and Duration Series (Fly Ash/Slag Ratio Constant at 50/50)

						AAC Mix Details					
Mix ID	% Na$_2$O	% SiO$_2$	w/b Ratio	Curing Temp	Curing Duration	Fly Ash (g)	Slag (g)	Na$_2$SiO$_3$ (g)	Extra Water Added (g)	Sodium Hydroxide (g)
AAC-60/24	8	8	0.38	60°C	24 h	500	500	301.89	165.55	74.29
AAC-85/24	8	8	0.38	85°C	24 h	500	500	301.89	165.55	74.29
AAC-60/48	8	8	0.38	60°C	48 h	500	500	301.89	165.55	74.29
AAC-85/48	8	8	0.38	85°C	48 h	500	500	301.89	165.55	74.29
AAC-60/72	8	8	0.38	60°C	72 h	500	500	301.89	165.55	74.29
AAC-85/72	8	8	0.38	85°C	72 h	500	500	301.89	165.55	74.29

TABLE 3.12

Details of AAC Mix for Curing Temperature and Duration Series (Fly Ash/Slag Ratio Constant at 40/60)

						AAC Mix Details				
Mix ID	% Na_2O	% SiO_2	w/b Ratio	Curing Temp	Curing Duration	Fly Ash (m)	Slag (m)	Na_2SiO_3 (g)	Extra Water Added (g)	Sodium Hydroxide (g)
AAC-60/24	8	8	0.38	60°C	24 h	400	600	301.89	165.55	74.29
AAC-85/24	8	8	0.38	85°C	24 h	400	600	301.89	165.55	74.29
AAC-60/48	8	8	0.38	60°C	48 h	400	600	301.89	165.55	74.29
AAC-85/48	8	8	0.38	85°C	48 h	400	600	301.89	165.55	74.29
AAC-60/72	8	8	0.38	60°C	72 h	400	600	301.89	165.55	74.29
AAC-85/72	8	8	0.38	85°C	72 h	400	600	301.89	165.55	74.29

TABLE 3.13
Details of AAC Mix for Water Curing Series

AAC Mix Details

Mix ID	% Na$_2$O	% SiO$_2$	w/b Ratio	Curing Duration	Fly Ash (g)	Slag (g)	Na$_2$SiO$_3$ (g)	Extra Water Added (g)	Sodium Hydroxide (g)
90–10	8	10	0.38	7/28 days	900	100	377.36	118.37	66.20
85–15	8	10	0.38	7/28 days	850	150	377.36	118.37	66.20
70–30	8	10	0.38	7/28 days	700	300	377.36	118.37	66.20
50–50	8	10	0.38	7/28 days	500	500	377.36	118.37	66.20
40–60	8	10	0.38	7/28 days	400	600	377.36	118.37	66.20

TABLE 3.14
Details of AAC Mix for Slag Series for Water Sorptivity

AAC Mix Details

Mix ID	% Na$_2$O	% SiO$_2$	w/b Ratio	Fly Ash (g)	Slag (m)	Na$_2$SiO$_3$ (m)	Extra Water Added (m)	Sodium Hydroxide (m)
100-0	8	10	0.38	1000	0	377.36	117.92	66.26
90-10	8	10	0.38	900	100	377.36	117.92	66.26
85-15	8	10	0.38	850	150	377.36	117.92	66.26
70-30	8	10	0.38	700	300	377.36	117.92	66.26

TABLE 3.15
Details of AAC Mix for Alkali Series for Water Sorptivity

	AAC Mix Details				AAC Mix Proportion			
Mix ID	% Na$_2$O in Activator*	% SiO$_2$ in Activator*	w/b Ratio*	Fly Ash (m)	GGBS (m)	Na$_2$SiO$_3$ (m)	Extra Water Added (m)	NaOH (m)
AAC-4%	4%	8.00%	0.38	700	300	301.89	177.52	21.08
AAC-6%	6.0%	8.00%	0.38	700	300	301.89	171.53	47.69
AAC-8%	8.0%	8.00%	0.38	700	300	301.89	165.55	74.29
AAC-10%	10.0%	8.00%	0.38	700	300	301.89	159.56	100.90

* By weight of binder material.

TABLE 3.16

Details of AAC Mix for Silica Series for Water Sorptivity

	AAC Mix Details			AAC Mix Proportion				
Mix ID	% Na$_2$O in Activator*	% SiO$_2$ in Activator*	w/b Ratio *	Fly Ash (g)	GGBS (g)	Na$_2$SiO$_3$ (g)	Extra Water Added (g)	NaOH (g)
AAC-4%	8.0%	4.00%	0.38	700.00	300.00	150.94	260.80	90.36
AAC-6%	8.0%	6.00%	0.38	700.00	300.00	226.42	213.18	82.32
AAC-8%	8.0%	8.00%	0.38	700.00	300.00	301.89	165.55	74.29
AAC-10%	8.0%	10.00%	0.38	700.00	300.00	377.36	117.92	66.26
AAC-12%	8.0%	12.00%	0.38	700.00	300.00	452.83	70.29	58.23
AAC-14%	8.0%	14.00%	0.38	700.00	300.00	528.30	22.67	50.20

* By weight of binder material.

TABLE 3.17

Details of AAC Mix for w/b Series for Water Sorptivity

	AAC Mix Details			AAC Mix Proportion				
Mix ID	% Na$_2$O in Activator*	% SiO$_2$ in Activator*	w/b Ratio*	Fly Ash (g)	GGBS (g)	Na$_2$SiO$_3$ (g)	Extra Water Added (g)	NaOH (g)
AAC-4%	8%	8%	0.35	700	300	377.36	87.92	66.26
AAC-6%	8%	8%	0.38	700	300	377.36	117.92	66.26
AAC-8%	8%	8%	0.41	700	300	377.36	147.92	66.26
AAC-10%	8%	8%	0.44	700	300	377.36	177.92	66.26
AAC-12%	8%	8%	0.47	700	300	377.36	207.92	66.26

* By weight of binder material.

3.6 RESULTS AND DISCUSSION

3.6.1 PREAMBLE

In this section, the results of the experimental study on engineering properties of AACs have been presented and scientifically interpreted, to have a deeper understanding of the influence of Na$_2$O, SiO$_2$, w/b ratio, curing temperature, curing duration, and fineness of particles on physico-mechanical properties and microstructure as well as their preparation and characterization. Subsequently, based on the results obtained from the experimental study, a mix design methodology and a prediction model based on regression analysis have been proposed to obtain compressive strength of the composite.

3.6.2 Engineering Properties of AACs in the Fresh State

3.6.2.1 Workability/Flow

3.6.2.1.1 Effect of Slag Content

Figure 3.20 represents the effect of slag content on the workability of AAC paste. Slag content was varied from 0% to 30% by weight of the total source material while keeping alkali content, silica content, and the w/b ratio of the mix constant at 8%, 10%, and 0.38%, respectively. Maximum and minimum average flow diameters were found to be 120 mm and 220 mm against slag content of 0% and 60%, respectively. It was observed that the flow diameter increased gradually from 0% to 60%. The percentage increase in flow diameter with respect to 0% slag content was observed to be 8.33%, 25%, 35.14%, 40%, and 45.45% with an increase in slag content 10%, 15%, 30%, 50%, and 60%, respectively. The relationship between flow and fly ash/slag ratio is noticed more or less linear. Increasing the slag content leads to a rise in the reactivity of the source material which results in more gel formation. Increasing the slag content increases the workability of AAC mix resulting in an increase in the flow diameter. The static and dynamic flow for different slag content is presented in Figure 3.21. There was 14.29%, 18.18%,

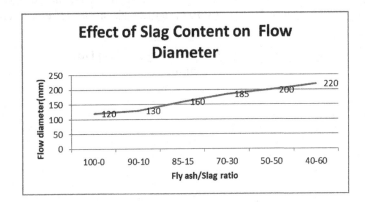

FIGURE 3.20 Effect of slag content on the flow diameter of AAC pastes.

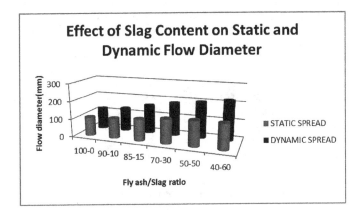

FIGURE 3.21 Effect of slag content on static and dynamic spread of AAC pastes.

33.33%, 37.04%, 42.86, and 57.17% increase in the dynamic flow diameter over static flow diameter against 0%, 10%, 15%, 30%, 50%, and 60% slag content, respectively.

3.6.2.1.2 Effect of Alkali Content

Figure 3.22 represents the effect of alkali content on the workability of AAC paste. Alkali content was varied from 4% to 12% by weight of (fly ash + slag) while keeping silica content and the w/b ratio constant at 10% and 0.38, respectively. Maximum and minimum average flow diameters were found to be 145 mm and 198 mm against alkali content of 12% and 8%, respectively. It was observed that, the flow diameter decreased suddenly for the change of alkali content from 8% to 10% and continues to decrease until 12% alkali content. The percentage increase in flow diameter with respect to 4% alkali content was observed to be 22.58%, 27.4%, and 12.9% for alkali content of 6%, 8%, and 10%, respectively. When the alkali content was increased beyond 10% up to 12%, the flow diameter dropped 3.23%, 6.45%. AAC paste mix with 12% Na_2O was very stiff while AAC mix with 8%Na_2O showed relatively higher workability. Increasing the alkali content up to the optimum percentage of 8% increases the flow of AAC mix. The static and dynamic flow for different percentages of alkali content is presented in Figure 3.23. There was 55%, 65.22%, 69.23%, 52.17%, 66.67% and 61.11% increase in dynamic flow diameter over static flow diameter against 4%, 6%, 8%,10%, and 12% alkali content, respectively.

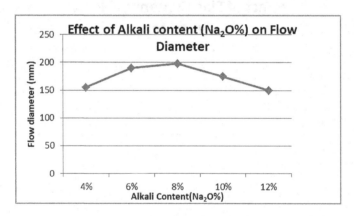

FIGURE 3.22 Effect of alkali content on workability of AAC pastes.

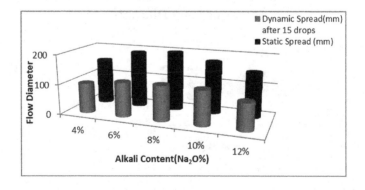

FIGURE 3.23 Effect of alkali content on static and dynamic spread of AAC pastes.

3.6.2.1.3 Effect of Silica Content

The Na_2O content and w/b ratio were kept constant equal to 8% and 0.38, respectively, and silica content (SiO_2) is changed from 4% to 14%. Figure 3.24 shows the effect of silicate content on flow diameter of AAC paste. It was observed from Figure 3.24 that flow diameter increases rapidly change of SiO_2 from 4% to 10%. Reduction in flow diameter was observed for the change of SiO_2 from 12% to 14%. Reduction of flow was due to an increase in sodium silicate content, because sodium silicate is more viscous than sodium hydroxide. The increase in silica content leads to a higher rate of dissolution which makes the paste stiff, reducing the flow diameter. The maximum average flow diameter of 220 mm and lowest flow diameter of 172 mm were observed against 10% SiO_2 and 14% SiO_2, respectively.

The percentage increase in flow diameter with respect to 4% silica content was observed as 1.11%, 10%, 22.22%, 2.78%, and −4.44% for 6%, 8%, 10%, 12%, and 14%, respectively. The static and dynamic flow for different percentages of silica content is presented in Figure 3.25. % An increase in dynamic flow diameter over

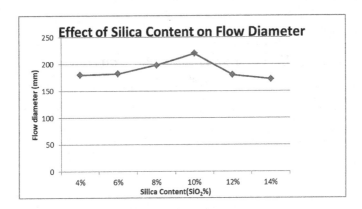

FIGURE 3.24 Effect of silica content on the flow diameter of AAC pastes.

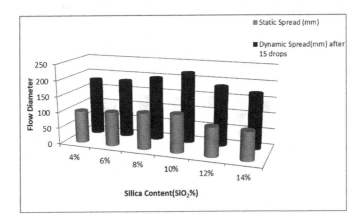

FIGURE 3.25 Effect of silica content on static and dynamic flow spread of AAC pastes.

respective static flow diameter was noticed 80%, 73.33%, 76.79%, 86.44%, 105.56%, and 95.45 against silica content of 6%, 8%, 10%, 12%, and 14%, respectively .

3.6.2.1.4 Effect of Water Content

The AAC mix composition was prepared by varying w/b ratio, keeping 8% Na_2O content and 10% SiO_2 content constant. The w/b ratio varied from 0.35 to 0.5. The effect of w/b ratio on the flow diameter of AAC paste is shown in Figures 3.26 and 3.27. It was noticed that the flow diameter increased gradually due to change of w/b ratio from 0.35 to 0.5. Maximum and minimum average flow diameter of 180 mm and more than 250 mm (flow occurred beyond the flow table) were observed for AAC mix having a w/b ratio of 0.47 and 0.5, respectively. The percentage increase in the flow diameter of the mix with respect to w/b ratio 0.35 was noticed to be 11.11%, 22.22%, 27.78%, 38.89%, and 38.89% for the mixes with a w/b ratio of 0.38, 0.41, 0.44, 0.47, and 0.50, respectively. The AAC mix with a w/b ratio of 0.38 showed moderate workability, whereas workability for a w/b ratio of 0.5 may be considered as high workability. The incorporation of extra water has thus shown good indication in effectively improving the flow of fresh AAC paste.

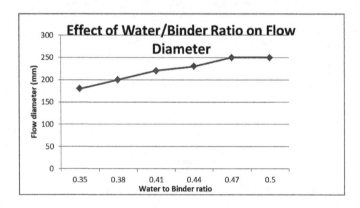

FIGURE 3.26   Effect of the w/b ratio on the flow diameter of AAC pastes.

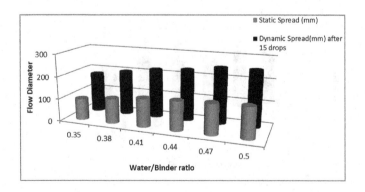

FIGURE 3.27   Effect of the w/b ratio on static and dynamic flow spread of AAC pastes.

3.6.2.1.5 Loss of Workability/Flow with Time

The loss of workability was measured with respect to relative reductions in flow diameter of AAC paste with time. The slag content of mix for test series 1 was 10%, 15%, 30%, 50% and 60%, and SiO_2 content of 10%, Na_2O content of 8%, and a w/b ratio of 0.38 were kept constant. The flow of different pastes was measured at 0 (initial), 10, 20, 30, 40, 50, and 60 minutes after mixing to determine the flow loss against the elapse of time. Figure 3.28 shows the effect of slag content on loss of workability/flow with time. It was observed that the rate of flow loss decreases with time. At the beginning of test, flow diameters of paste were in the range of 130–220 mm with the highest and the lowest values recorded by the mix specimens 40–60 and 90–10, respectively. However at 60 minutes, the highest percentage decrease of 52.28% was recorded by the mix 40–60. It was noticed that the initial flow diameter was seen to increase with addition of slag, but the percentage decrease of the diameter with time was also found to increase with a greater amount of slag in the source material. The alkali content of mix for test series 2 was 4%, 6%, 8%, 10%, and 12%, and SiO_2 content of 8% and a w/b ratio of 0.38 were kept constant. The flow of different pastes was measured at 0 (initial), 10, 20, 30, 40, 50, and 60 minutes after mixing to determine the flow loss against the elapse of time. Figure 3.29

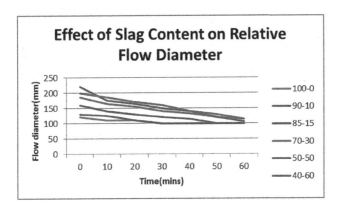

FIGURE 3.28 Effect of slag content on the flow diameter with time of AAC pastes.

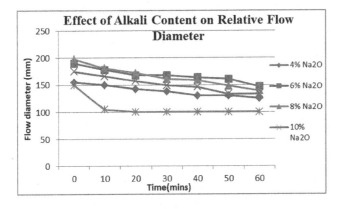

FIGURE 3.29 Effect of alkali content on the flow diameter with time of AAC pastes.

shows the effect of alkali content on loss of workability/flow with time. It is noticed that the rate of flow loss decreases with time. At the beginning of test, flow diameters of paste were in the range of 175–190 mm. At 60 minutes, the diameter of the mixes with 8% Na_2O and 10% Na_2O was noticed to remain almost unchanged. At 30 minutes, however, flow diameters of all the pastes are observed to be in the range of 150–170 mm except the mix A5 (12% Na_2O). However at the elapse of time at 60 minutes, flow diameters of the pastes having alkali content of 4%, 6%, and 8% were in the range of 125–147 mm. However, the paste with 12% alkali content had a constant diameter of 100 mm after 20 minutes. Faster dissolution due to the increase in alkali content causes the mix with 12% Na_2O content to flash set and become unworkable. It is observed that mixes having alkali content of 4%, 6%, 8%, and 10% have a reasonable workability/flow. The silica content of mix for test series 3 was 4%, 6%, 8%, 10%,12%, and 14%, and SiO_2 content of 8% and a w/b ratio of 0.38 were kept constant . For test series 3, the flow diameter varied from 180 mm to 220 mm (Figure 3.30a). The flow diameter decreased substantially after

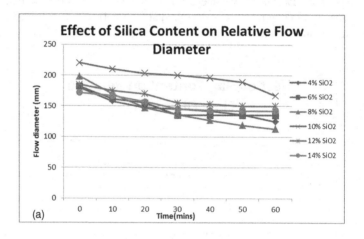

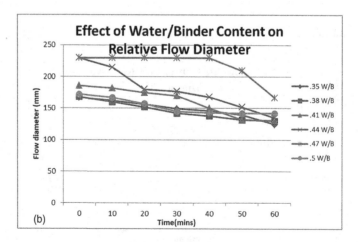

FIGURE 3.30   (a-b) Effect of w/b ratio on the flow diameter with time of AAC pastes.

10 minutes and the values were between 158 mm and 210 mm. The highest and lowest flow diameters at 0 minutes and 60 minutes were 167 mm and 125 mm for B4 and B1 mix . The water content of mix for test series 4 was 0.35, 0.38, 0.41, 0.44, 0.47, and 0.5 and SiO_2 content of 10% and Na_2O content of 8% were kept constant. For test series 4, the flow diameter varied from 167 mm to 230 mm. The highest and lowest flow diameters at 60 minutes were 230 mm and 125 mm for C6 and C1 mix, respectively (Figure 3.30b).

3.6.2.2 Setting Time

The initial and final setting time of AAC paste was determined at room temperature using Vicat apparatus. Four series of tests were conducted. In the first series, the slag content was varied from 10% to 60% by keeping constant Na_2O 8%, SiO_2 8%, and w/b ratio 0.38. In the second series, the effect of alkali content was studied by varying the $\%Na_2O$ from 4% to 12% by keeping silicate content 8% and a w/b ratio of 0.38 constant. In the third series, alkali content (8%) and w/b ratio of 0.38 was kept constant and silicate content was varied from 0% to 14%. In the fourth series, the w/b ratio was varied from 0.35 to 0.5 by keeping Na_2O and SiO_2 content equal to 8%.

3.6.2.2.1 Effect of Slag Content

Figure 3.31 shows the effect of slag content on the setting time of AAC paste. AAC pastes were prepared by varying the slag content from 10% to 60%. The SiO_2, Na_2O, and w/b ratio was 8%, 8%, and 0.38, respectively. From figure 3.31, it is observed that the initial setting time is decreasing with the increase of slag. The initial setting time of slag-based AAC paste was recorded as 180, 172, 164, 145, and 120 minutes and final setting time as 200, 193, 188, 159, and 130 minutes for 10%, 15%, 30%, 50%, and 60% of slag content, respectively. As the slag amount increased, there was a substantial decrease in the time difference between initial and final setting time. This can be explained from the fact that the slag particles serve as additional nucleation centers, thus increasing the intensity of the polymerization as whole. As the dissolution, gelation, solidification and polycondensation occur simultaneously in a

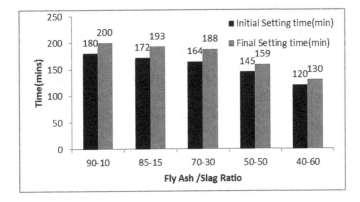

FIGURE 3.31 Effect of slag content on initial and final setting time of AACs.

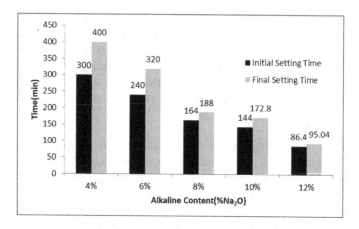

FIGURE 3.32 Effect of alkaline content on initial and final setting time.

polymeric reaction, if the intensity of the reaction process increases a few particles will be at an advance stage of polymerization than the others thus reducing the time between the initial and the final setting time.

3.6.2.2.2 Effect of Alkali Content

Figure 3.32 shows the effect of alkali content on the setting time of AAC paste. AAC pastes were prepared by varying the alkali content from 4% to 12%. The SiO_2 and w/b ratio was kept constant at 10% and 0.38, respectively. From Figure 3.32, it was observed that the increase in the addition of NaOH made a considerable change to the initial as well as final setting time. However, the change was more prevalent in the lowering of the difference between the final and initial setting time. The initial setting time of slag-based AAC paste was recorded as 300, 240, 164, 144, and 86.4 minutes and final setting time as 400, 320, 188, 172.8, and 95.04 minutes for 4%, 6%, 8%, 10%, and 12% of alkali content, respectively.

3.6.2.2.3 Effect of Silica Content

Figure 3.33 shows the effect of silica content on setting time of AAC paste. It was observed that the increase in silicate content results in a decrease in initial and final setting of AAC paste. There was a decrease in initial setting time at a rate of 5.35%, 18.67%, 39.72%, 51.54, and 62.81% (with respect to 4% SiO_2 content) for 6%, 8%, 10%, 12%, and 14% SiO_2 content, respectively. The final setting time was decreased at a rate of 11.31%, 28.8%, 55.7%, 72.03%, and 84.96% with respect to 4% SiO_2 content for 6%, 8%, 10%, 12%, and 14% of silicate content. Dissolvable silicate phases are known to increase the rate of dissolution as accelerated rate of dissolution leads to an earlier polycondensation reaction, which leads to early stiffening of the gel phases.

3.6.2.2.4 Effect of Water Content

Figure 3.34 shows the effect of w/b ratio on setting time of AAC paste. It was observed that the increase in the w/b ratio results in an increase in initial and final setting of AAC paste. There was an increase in initial setting time at a rate of 10.81%, 21.62%, 39.86%, 64.86%, and 101.35% (with respect to 4% SiO_2 content) for a w/b ratio of

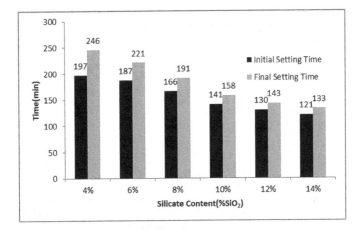

FIGURE 3.33 Effect of silicate content on initial and final setting time of AAC paste.

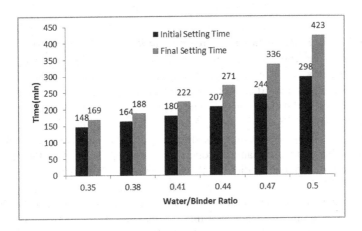

FIGURE 3.34 Effect of w/b ratio on initial and final setting time of AAC paste.

0.35, 0.38, 0.41, 0.44, 0.47, and 0.5, respectively. The final setting time increased at a rate of 27.03%, 50%, 83.11%, 127.03%, and 185.81% (with respect to 0.35 w/b ratio) for a w/b content of 0.38, 0.41, 0.44, 0.47, and 0.5, respectively. In the polymerization reaction, the water does not participate in the reaction directly but instead acts as a transportation medium. Here, the water–binder ratio above 0.38 does not enhance the setting time of AACs but nonetheless it increases the fluidity of the mix.

3.6.3 HARDENED PROPERTIES OF BLENDED AAC EFFECT OF SYNTHESIS PARAMETERS ON AACS CONTAINING FLY ASH AND SLAG

3.6.3.1 Effect of Variation of Slag

3.6.3.1.1 Compressive Strength

The compressive strength of the specimens is seen to increase with addition of slag content. 19.99% was the increase in the compressive strength of the mix 90-10 over

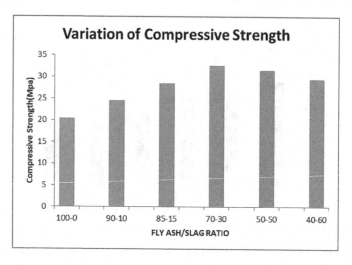

FIGURE 3.35   Effect of variation of slag content on compressive strength of blended AAC.

mix 100-0. Mix 85-15 reports an increase of 15.97% over the mix 90-10 .The mix 70-30 sees an increase of compressive strength of 14.75% over the mix 85-15 . However, the mix 50-50 and mix 40-60 report a decrease of −3.428% and −6.33% over mix 70-30 and mix 50-50, respectively (Figure 3.35) . The increase of strength can be attributed to the more reactive nature of slag particles. The formation of C-S-H gel (calcium silicate hydrate) and C-A-S-H (calcium alumina-silicate hydrate gel along with N-A-S-H (sodium alumina-silicate hydrate) gel has been reported to have a positive effect on the mechanical properties of AACs [21–23]. Also Puertas, Provis et al have reported in their studies that the addition of slag leads to high early strength. This seems to be in sync with results of Kumar et al. [23] who also reported an increase in compressive strength with addition of slag into a fly ash/slag blended composite. Yip et al. also suggested that the increase in the compressive strength could be due to the coexistence of both C-S-H and geopolymeric gel [, 109]. The presence of Ca could also cause degradation of the geopolymeric gel at high pH, and excess Ca in the system could lead to the formation of a highly disordered and unstable geopolymeric gel. Calcium ions can preferentially form (Ca, K)-A-S-H as the pH of the system lowers in the later age [110]. This could be the reason for the degradation of mechanical properties in mix 50-50 and mix 40-60 in the current study. Yip et al. also reported that the presence of Ca could disrupt the three-dimensional (3-D) ordering of geopolymers. Another study by I. García-Lodeiro et al, however, reports that Ca ion substitution into a geopolymeric gel network may not cause disordering of the structure but can result in changes of chemical composition [111]. Marjanović et al. reported that the increase in the GGBS content led to a lesser w/b ratio which in turn resulted in greater compressive strength with the 25% FA–75% BFS mix reporting the highest values [112]. Wardhono et al also reported an increase in compressive strength with slag addition with the mix with 50% slag giving the highest strength [113]. Zhang et al reported a decrease in the strength value in a fly ash, slag-based composite subjected to ambient temperature curing with slag percentage more than 30% which is in agreement with this study [114]. Further discussions about the other engineering properties are made in the following section.

3.6.3.1.2 Bulk Density, Water Absorption, Apparent Porosity, and UPV

The bulk density for the mix 100-0 with fly ash as the sole precursor is seen to have the lowest bulk density value at 1637.06 kg/m^3 (Figure 3.36).The value increases to 1761.66 kg/m^3 for mix 90-10 with 10% addition of slag by weight. The bulk density value for mix 85-15 is a bit higher at 1798.83 kg/m^3. It continues to increase for mix 70-30, with a value of 1916.87 kg/m^3. However, the mix 50-50 and mix 40-60 show a decrease in bulk density values from the previous mixes at 1699.72 kg/m^3 and 1593.74 kg/m^3, respectively. The results clearly show that the bulk density of the specimens increases with the addition of slag up to an optimum level of 30% .This can be explained by the fact that the incorporation of slag which is calcium-rich in nature leads to the formation of a denser microstructure which is further confirmed by images as discussed in Section 3.6.3.1.4. Though slag has a more reactive nature, which leads to high early strength, the extra charge balancing aluminum ions of fly ash contribute to the long-term nature of the gel structure increasing its rigidity in the process. The above-made arguments about the gel structure of the alkali-activated fly ash/slag AACs can also be corroborated by the apparent porosity and water absorption results. The apparent porosity of the samples decreases with the increase of slag content up to the optimum level of 30% (Figure 3.37). The lowest apparent porosity value was found to be lowest at 14.61% for the 70-30 mix and the highest value was found for mix 100-0 at 22.92%. This can be attributed to the fact that the high reactivity of the slag particles leads to a higher amount of gel formation. The flow diameter has also been observed to increase with slag addition though it did not follow the trend of the other findings and continued to increase up to the mix with the highest amount of slag. The increase of the flow leads to smooth pouring, compaction, and casting of the mixes. The mixes with more than 30% slag did not need mechanical vibration. Also the spherical nature of the fly ash particles leads to the formation of a dense microstructure with lesser amount of pores. The UPV data also

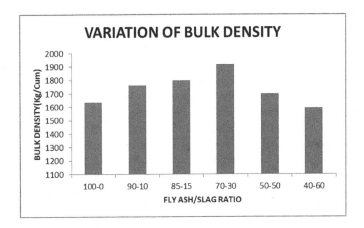

FIGURE 3.36   Effect of variation of slag content on bulk density of fly ash–slag-based AACs.

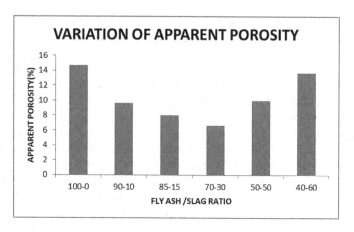

FIGURE 3.37   Effect of variation of slag content on apparent porosity of fly ash–slag-based AACs.

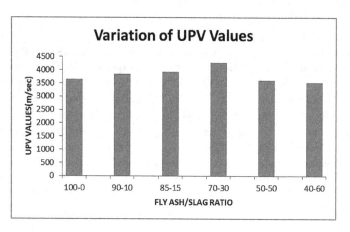

FIGURE 3.38   Effect of variation of slag content on UPV values of fly ash–slag-based AACs.

points to this fact. The UPV values have also been found to increase with the higher slag percentage up to the mix 70-30 (Figure 3.38). The UPV values of the mix 70-30 were the maximum, and the phenomenon is in accordance with previous studies involving the microstructure of fly ash–slag blends where it has been reported that a major amount of gel formation involves cross linkages of C-S-H and N-A-S-H gel leading to the formation of a hybrid, rigid 3D tetrahedral gel network [21].The water absorption values followed the same trend as discussed above (Figure 3.39). The dissolution of silicon and aluminum species from fly ash is very slow in general. On addition of slag, the dissolution rate increases which contributes to the improvement of the gel network and subsequently the engineering properties. The changes in the gel network can be further explained with the XRD, FTIR, and field emission scanning electron microscopy (FESEM)/EDX analysis discussed in the following sections.

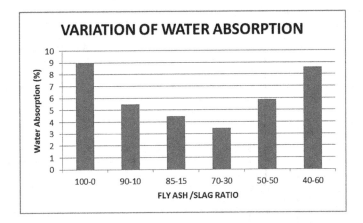

FIGURE 3.39   Effect of variation of slag on water absorption values of fly ash–slag-based AACs.

3.6.3.1.3 Mineralogical Investigation by XRD

Q- Quartz
Al-C-S-H-Aluminum-calcium-silicate-hydrate
M- Mullite
C-Calcite
C-S-H-Calcium Silicate Hydrate
S-Sillimanite

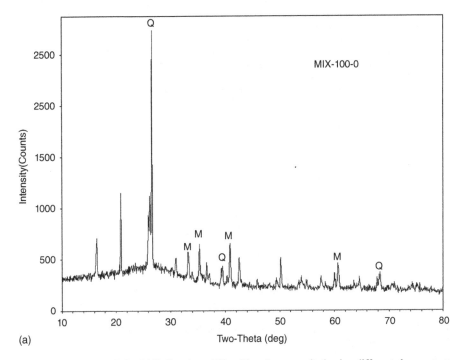

(a)

FIGURE 3.40   (a) XRD of Alkali activated Slag-Fly ash composite having different slag content.

(*Continued*)

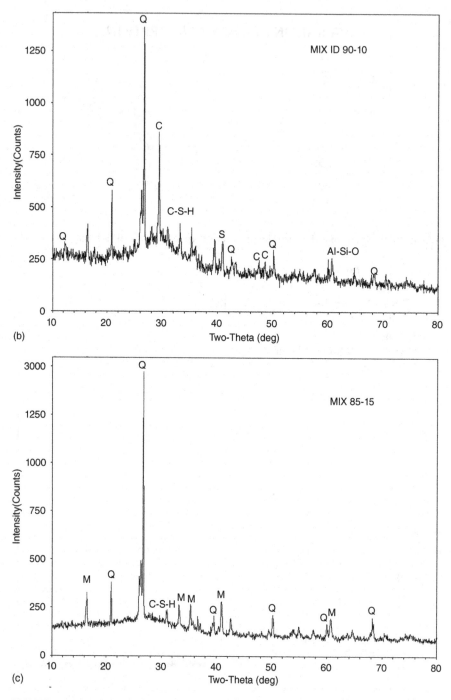

FIGURE 3.40 (CONTINUED)   (b–c) XRD of Alkali activated Slag-Fly ash composite having different slag content.

(Continued)

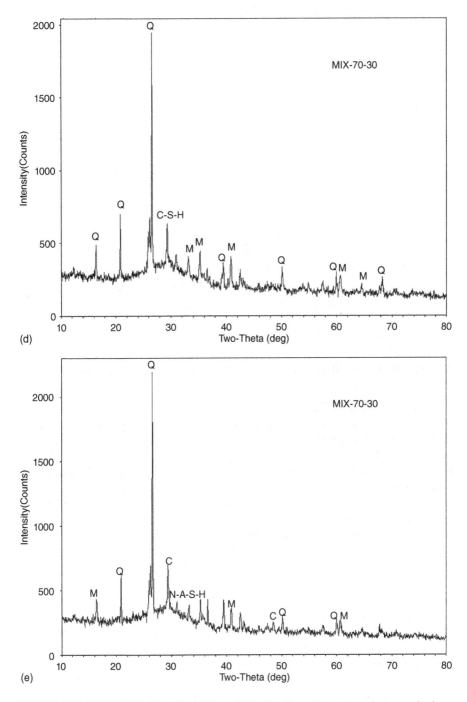

FIGURE 3.40 (CONTINUED) (d–e) XRD of Alkali activated Slag-Fly ash composite having different slag content.

(*Continued*)

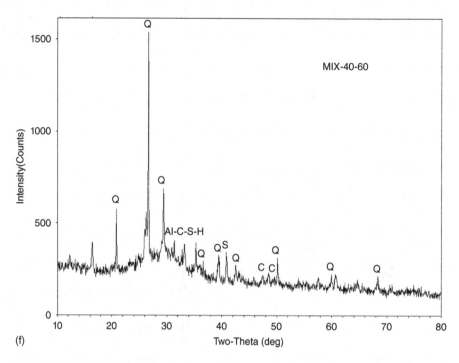

(f)

FIGURE 3.40 (CONTINUED)   (f) XRD of Alkali activated Slag-Fly ash composite having different slag content.

Figure 3.40a–f provided XRD data of the alkali-activated slag–fly ash composites. Specimen 100-0 shows peaks of mainly quartz and mullite compounds which can be termed as N-A-S-H gel which is in tune with other findings associated with alkali activation of fly ash-based AACs. Identifying alumina–silicate gel accurately is a problem as it is highly amorphous in nature; thus the peaks of crystalline quartz and mullite may be used as an indicator to comment on the intensity of alumina–silicate gel. With addition of slag into the mix 90-10, secondary peaks of calcite and C-S-H gel are to be seen. These peaks appear along with the primary peaks of quartz and mullite. For the mix 85-15, the same trend was noticed regarding the peaks identified as mullite, quartz, calcite, and C-S-H gel. In the mix 70-30, the intensity of the peak due to quartz is less than that of the mix 85-15.The intensity of the C-S-H gel is greater than that of the mix 85-15. These data can be used to argue that as the slag percentage increases, there is an increase in calcium-based compounds which essentially functions as a supplementary gel network to the major binding gel network consisting of alumina–silicate compounds. This is supported by the fact that even though there is an increase in peak intensity of calcium-based compounds, the highest peaks are still that of alumina–silicate compounds. In the present study, as the curing temperature was set at 85°C, a higher degree of polymerization of the fly ash particles took place. But as the slag content increased to 50% and 60% in mix 50-50 and mix 40-60, the mechanical properties were found to decrease .This can be partially explained by the fact that peak intensities for crystalline quartz lessen substantially thus indicating the decrease of alumina–silicate gel which acts as the primary gel network in the other mixes. Thus, a point can be made that the calcium compounds are responsible for enhancing initial reactivity of the mix but under the effect of high temperature curing, it plays the role of a secondary gel network.

3.6.3.1.4 SEM/ EDX

Referring to SEM images (Figure 3.41 a-f, where a,b,c,d,e and f represent 100-0,85-15,70-30,50-50 and 40-60 specimens respectively.) it may be noted that there has been a gradual change in the microstructure with addition of slag. The image of mix 100-0

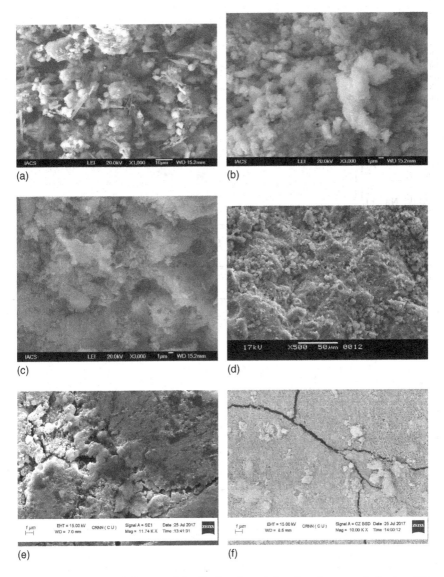

(a)

(b)

(c)

(d)

(e)

(f)

FIGURE 3.41 (a–f) SEM image of the alkali-activated slag–fly ash composite having different slag contents. (*Continued*)

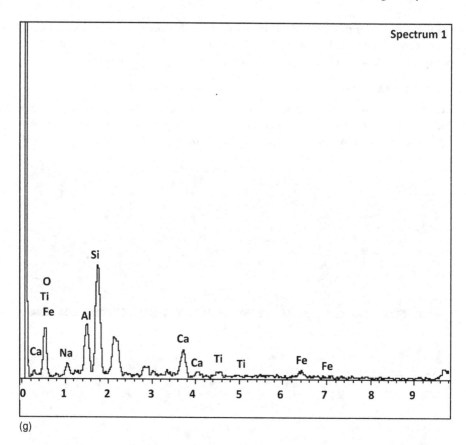

(g)

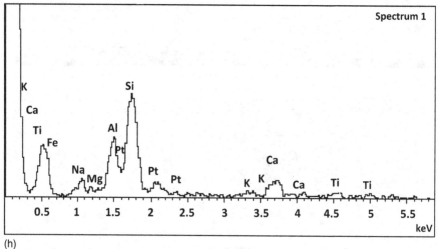

(h)

FIGURE 3.41 (CONTINUED)   (g–h) EDX of the alkali-activated slag–fly ash composite having different slag contents.

(Continued)

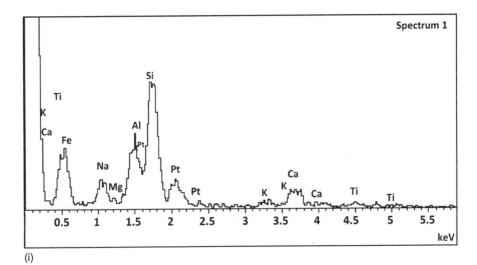

(i)

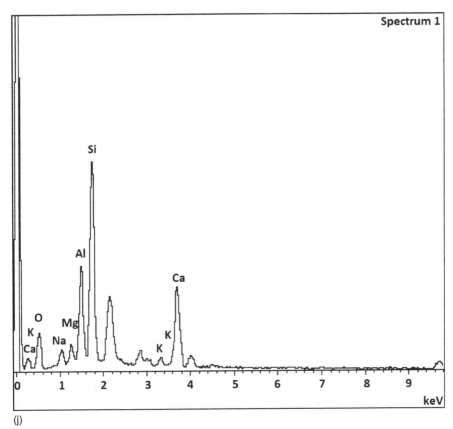

(j)

FIGURE 3.41 (CONTINUED)   (i–j) EDX of the alkali-activated slag–fly ash composite having different slag contents.

(Continued)

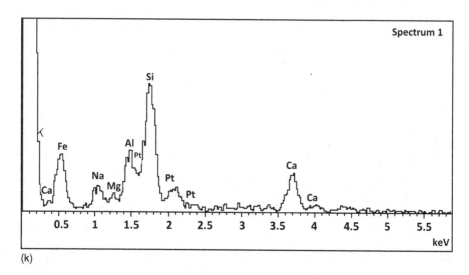

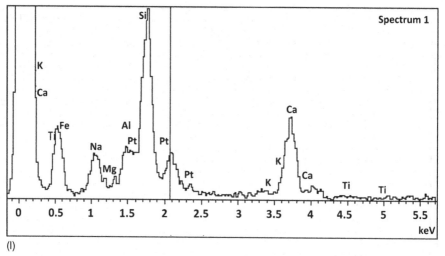

FIGURE 3.41 (CONTINUED) (k–l) SEM image of the alkali-activated slag–fly ash composite having different slag contents.

shows the existence of large number of unreacted spherical fly ash particles which do not contribute to the rigidity of the gel network and also hinder the strength gaining process. The mix 90-10 image also shows a few unreacted fly ash particles but the gel formation is noticeably more. The fly ash particles have been coated with the gel. The reaction products can be classified as mainly alumina-silicate in nature as the EDX results (Fig 3.41 g-l,where the g,h,i,j,k and l represent 100-0,90-10,85-15, 70-30,50-50,40-60 respectively.) are showing much higher peaks of silica and aluminum than calcium. The mix 85-15 is seen to more amorphous in nature with very few unreacted fly ash particles, and the slag particles angular in nature are assumed to fully participate in the gel formation as no angular particle can be seen. The pores present in the structure can also be noticed. The mix 70-30 SEM image shows structure with a reduction in pores. A few unreacted fly ash particles can still be noticed but these have been covered with reaction products. In the EDX analysis, there is a spurt in the peak of calcium which is in tandem with the increase in slag concentration. In the mix 50-50, the separation of phases between the N-A-S-H and C-A-S-H gel is seen clearly, which in results in lesser amount of cross linkages and lessens the cohesive nature of the mix to an extent. The fly ash particles seem to have participated in the formation of the reaction products which can also be justified by the fact that there is lesser amount of fly ash in the mix 50-50 to participate in the dissolution process and more amount of slag, which accelerates the dissolution process showing an increase in the percentage of calcium. In the mix 40-60 image, it is noticed that there are more pore formations than the mix 50-50 . Similar findings in case of apparent porosity were observed.

3.6.3.1.5. FTIR Analysis

Intensity of the broad band between 3200 cm^{-1} and 3600 cm^{-1} appears to increase with the increase in percentage of slag content which may be due to the increased early formation of hydration products. Absorption bands from 1670 to 1690 cm^{-1} are assigned to bending vibration O-H groups of the hydrated product and the band shows an increase in intensity centered at 1671 cm^{-1} (Figure 3.42). The wavenumbers associated with Si-O stretching vibrations are seen shifting to lower wave numbers which can be linked to the fact that the simultaneous activation of fly ash and slag leads to more crosslinked N-C-A-S-H gel. The band connected with the formation of the main binder gel, for example, C-A-S-H gel and N-A-S-H gel in alkali-activated materials has been mainly found in the range of wave numbers from 950 to 1100 cm^{-1} [114]. In

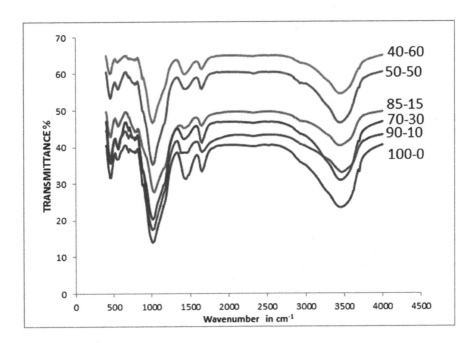

FIGURE 3.42 FTIR spectra of alkali-activated fly ash–slag paste specimens having different fly ash/slag ratios.

the present study, the main binder gel has been noticed between 1031 and 1049 cm⁻¹ for various mixes having different slag contents (Figure 3.42). The wave numbers indicate clearly the formation of a hybrid 3-D calcium alumina–silicate gel structure.

3.6.3.1.6 TGA/DTA

Referring to the above TGA profiles (Figure 3.43) it may be noted that the highest percentage of mass loss is 15.12% for the mix 40-60 (Figure 3.44).The mix with fly ash as the sole precursor had a mass loss of 8.92%, and the mix 90-10 had a mass loss of 12.68%. Other mixes showed a trend of increase of mass loss percentage with the increase of slag content, that is, mix 85-15, 70-30, and 50-50 recording 13.12%, 13.24%, and 14.95%, respectively. Physically bound water in the gel structure is released below temperature of 115°C .The samples show a decrease in peak temperature values with the increase in slag content. As fly ash particles are spherical and of higher fineness, less water is trapped in the mix. In reported studies on pore size distribution of fly ash–slag composites, it has be seen that the amount of macro-pores (size 25—5000 nm) in slag-based composites is more compared to fly ash-based composites (only fly ash) [115,116]. As macro-pores hold a significant amount of free

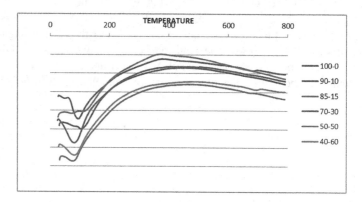

FIGURE 3.43   TGA/DTA–Graphical representation of mass loss peaks of alkali-activated fly ash–slag AAC paste of different fly ash/slag ratios.

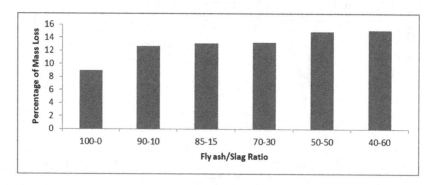

FIGURE 3.44   Percentage of mass loss of alkali-activated fly ash–slag pastes of different fly ash/slag ratios.

water, the slag-based composites are reporting peak temperature intensities at lower values. There are also other peak intensities reported between 550°C and 600°C.

3.6.3.1.7 Pore Structure Study by MIP

The pore characteristics of AAC paste specimens have been investigated with the help of MIP. In this test, mercury has been forced, using pressure into the pores of a material. The pore size distribution was determined from the volume intruded at each pressure increment. Total porosity was determined from the total volume intruded. The pore properties of the specimens have been plotted as the cumulative volume of mercury intruded versus pore diameter. Figure 3.45 shows the result of the cumulative pore size distribution by MIP for specimens 100-0, 85-15, 70-30, 50-50, and 40-60. It was observed that more or less when the percentage of slag percentage increases, the cumulative mercury intrusion decreases. MIP results show that most of the pores are in between 0.01 and 1 µm (Figures 3.45 and 3.46).

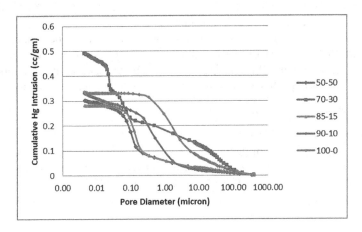

FIGURE 3.45   Cumulative pore size distribution for blended AAC paste.

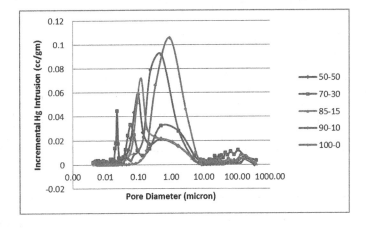

FIGURE 3.46   Incremental Hg volume intruded with respect to the diameter of pores.

3.6.3.1.8 Summary

1) Addition of slag up to an optimum level of 30% increases the mechanical properties, for example, compressive strength, bulk density, apparent porosity, and water absorption values of thermally cured fly ash-based AACs. The presence of excess positively charged calcium ions seems to be negatively affecting the gel network .However, the flow diameter of the mix increased almost linearly with the addition of slag.

2) Variation of slag content causes changes to the microstructure of the AACs. Addition of slag leads to the formation of two distinct gel phases, namely, N-A-S-H or geopolymeric gel and C-A-S-H gel. XRD analysis also points to the presence of a cross-linked gel phase N-C-A-S-H gel as well. The N-A-S-H gel primarily functioned as the major load-bearing gel phase. In general, reaction products bore more resemblance to alkali-activated fly ash composites rather than alkali-activated slag-based composites. The main reason behind this can be attributed to use of a higher curing temperature of 85°C which is more conducive to the polymerization of fly ash.

3) TGA/DTA pointed to the fact that the presence of higher percentage of slag leads to the presence of freely available water thus causing greater mass loss of test samples with a higher slag content with the increase of the temperature level. The sample having fly ash as the sole precursor reported lowest mass loss.

3.6.3.2 Effect of Alkali Content

3.6.3.2.1 Compressive Strength

The compressive strength of the mixes is seen to increase with the increase of alkali content (Figure 3.47). There is an exponential increase of 154.55% in the compressive strength value of A2 - ($6\%Na_2O$) mix over A1($4\%Na_2O$) mix. A3 ($8\%Na_2O$) mix reports an increase of 15.97% over the A2 mix. A4 ($10\%Na_2O$) mix reports a

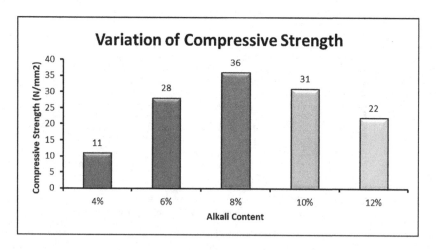

FIGURE 3.47   Effect of variation of alkali content (%Na_2O) on compressive strength of fly ash–slag-based AACs.

decrease of compressive strength of 6.06% over A3 mix. However, A5 (12%Na$_2$O) mix reports a decrease in strength of 37.149%. A higher amount of hydroxyl ion concentration in the solution quickens the dissolution of Si^{4+} and Al^{3+} from precursor materials and acts as a reaction catalyst, while Na+ acts as a charge-balancing cation in the structure [117,118]. However, high NaOH concentration creates an environment of higher pH which hinders the leaching rate of Ca$^+$ ions. Thus at an optimum alkali concentration, in this case 8%, Ca also contributes in the formation of a disordered structure with weak Al-O-Al bonds [49]. The presence of adequate Ca$^+$ ions in the gel framework promotes the formation of a multiphase system having both calcium silicate hydrate (C–A–S–H) and geopolymer (N–A–S–H) gels [119,120]. Studies have also shown that calcium content in alkali-activated solid waste materials functions as extra sites of nucleation [121,122]. One of the reasons for the dissolution rate increasing with addition of alkali content (Van Jaarsveld and Van Deventer, 1999) is that Si^{4+} and Al^{3+} species are released from the fly ash particle surface and into the solution increasing the thickness of the solution which in turn is the first step toward gelation stage of geopolymer. Thus from Van Jaarsveld and Van Deventer's findings, they suggested that leaching was, therefore, dependent on alkali solution concentration. U. Rattanasak et al studied the actual leaching rates of Si^{4+} and Al^{3+} ions and reported that for a 10M NaOH solution, the concentration of Si^{4+} ions is approximately 600 ppm .The rate of dissolution was accelerated when the molarity of the solution was increased to 10M [79]. However, the increase in rate of dissolution with the increase in alkaline concentration was not linear in nature. The concentration decreased when the molarity of the solution was increased to 10M which led to the coagulation of silica species [122], which is noticed to be similar. The compressive strength decreased when the percentage of NaOH crosses 8% by the weight of the source material. The concentration of Si^{4+} decreased to 200 and 260 ppm for solutions of 5M NaOH and 15 NaOH. Excess of OH$^-$ ions also causes a setback to the process of polycondensation which in turn has as a negative effect on the compressive strength [123]. Similar observations have been made in the present study.

3.6.3.2.2 Bulk Density, Water Absorption, Apparent Porosity, and UPV

The bulk density for the A1 (4% Na$_2$O) mix is seen to have the lowest bulk density value at 1700.27 kg/m^3 (Figure 3.48). The value increases to 1745.95 kg/m^3 for A2 (6% Na$_2$O) mix. The bulk density value for A3 (8% Na$_2$O) mix is highest at 1842.26 kg/m^3. It decreases slightly for A4 (10% Na$_2$O) mix with 1837.95 kg/m^3 but it registers a steep drop for the A5 (12% Na$_2$O) mix with a value of 1622 kg/m^3. Higher bulk density signifies closer packing of the molecules which in this case is supported through the images of the microstructure which are in sync with the bulk density. The closer packing about the gel structure of the alkali-activated fly ash–slag AACs can also be corroborated by the apparent porosity and water absorption results. The apparent porosity of the specimens decreases with the increase of alkali content up to the optimum level of 8% (Figures 3.49 and 3.50). The lowest apparent porosity value was found to be at 4.71% for the A3 mix and the highest value was found for A5 mix at 14.23%.Water absorption percentages for the different mixes followed the same trend with the A1 and A4 mix recording the highest and lowest values,

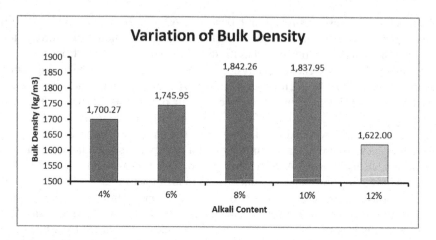

FIGURE 3.48 Effect of variation of alkali content (%Na$_2$O) on bulk density of fly ash–slag-based-AACs.

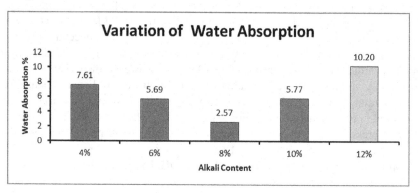

FIGURE 3.49 Effect of variation of alkali content (%Na$_2$O) on water absorption of fly ash–slag-based AACs.

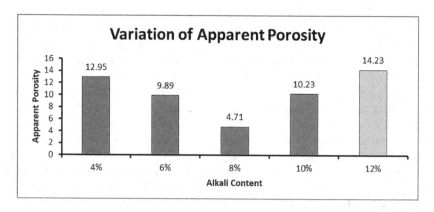

FIGURE 3.50 Effect of variation of alkali content (%Na$_2$O) on apparent porosity of fly ash–slag-based AACs.

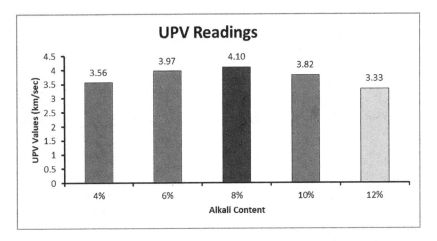

FIGURE 3.51 Effect of variation of alkali content (%Na$_2$O) on UPV of fly ash–slag-based AACs.

respectively. The changes in the abovementioned physical properties can be corroborated with UPV values as well. The UPV values of the A1, A2, A3, A4, and A5 mix are 3.56 km/s, 3.97 km/s, 4.10 km/s, 3.82 km/s, and 3.33 km/s, respectively (Figure 3.51). At low alkalinity, neither the alumina and silica species from fly ash nor the calcium species from slag are sufficiently dissolved thus leading to lower physical properties. From the micrographs, it can be seen that the increase in the concentration of sodium hydroxide leads to the formation of more homogeneous AACs but only up to the optimum alkali content of 8%. The calcium present in the composites acts as a seeding or precipitating element. The EDX results also show that the calcium content rises up to A3 sample but starts decreasing thereafter thus reiterating the statement that high alkalinity obstructs the calcium from participating in the polymerization reaction. The results signify that alkali content up to 8% is beneficial for the formation of a more cohesive microstructure in fly ash-based AACs. The changes in the gel network can be further explained with the FESEM images given in the FESEM/EDX analysis in the subsequent section.

3.6.3.2.3 Mineralogical Investigation by XRD Analysis

XRD data of the A1 mix as shown in Figure 3.52, suggest that the basic ingredients of fly ash such as quartz and mullite along with calcite due to the slag are present in it. However, the data did not point to the presence of sialate and polysialate atoms that are the main products of polymerization. In the A2 mix, it was noticed that in addition to the peaks of quartz and mullite, there were peaks denoting sillimanite, an alumino silicate compound which signified an improved level of polymerization. In the A3 mix, it can be seen that an alumino-calcium-silicate hydrate gel phase can be seen in the peaks. This phase is the result of a high amount of cross linkage between the polymeric gel phase and calcium silicate hydrate gel phase, for example, a product of the alkali activation of slag. As the percentage of alkali content increases in the A4 mix, the order of cross-linkage between the aluminum, silica, and the calcium

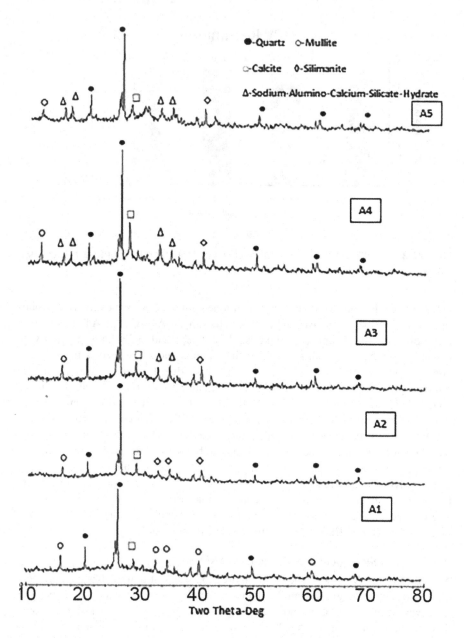

FIGURE 3.52 Effect of variation of Na$_2$O content on XRD graphs of fly ash–slag-based AACs.

ions increases. As a result, there are more peaks of cross-linked alumina calcium silicate gel in the A4 mix than A3 mix. Analyzing the abovementioned peak diffraction data, it can be seen that the percentage of sodium hydroxide is an integral factor controlling the process of polymerization. Polymerization of alkali-activated materials can be divided into four stages (i) dissolution of Si and Al from the solid alumino–silicate materials in a strong alkaline aqueous solution, (ii) formation of Si and Si-Al oligomers in the aqueous phase, (iii) polycondensation of oligomers to form a 3-D aluminosilicate framework, and (iv) bonding of the undissolved solid particles into the geopolymeric framework and hardening of the whole system into a final solid polymeric structure. By closely monitoring the XRD data, it can be said that in the A1 mix due to inadequate NaOH concentration, dissolution of the Si and Al ions did not take place properly thus hampering the progress of the subsequent stages. As the percentage of NaOH started increasing and reached an adequate amount in the A3 and A4 mix, the first stage, for example, dissolution of Si and Al ions was completed properly.

3.6.3.2.4 Microstructure Study by SEM/EDX

As discussed in the previous section regarding XRD analysis, the concentration of NaOH ions in the activator solution has an effect on the progress of polymerization at various stages. This is clearly reflected in the SEM images (Figure 3.41a–e,where a,b,c,d and e represent 4% Na_2O,6% Na_2O,8% Na_2O,10% Na_2O and 12% Na_2O specimens respectively). SEM images of the A1 (4% Na_2O) mix clearly point to the fact that the gel structure is disjointed in nature leaving multiple voids in the structure, a few unreacted fly ash particles can also be seen. The lack of adequate OH^- ions resulted in the incomplete dissolution of Si, Al, and Ca species; thus a lot of inactivated fly ash and slag particles are seen. In the A2(6% Na_2O) mix, an improved and more cohesive microstructure with a lesser number of voids is noticed. However, the gel matrix has still not been properly formed, and unreacted fly ash slag particles still remain. In A3 (8% Na_2O) mix, it is seen that relative to other samples more gel formation has taken place and a few products of precipitation are seen. The calcium content also rises in this mix compared to A1(4% Na_2O) and A2(6% Na_2O). As the alkali content is increased up to 10% (A4 sample), cracks are seen to develop. From the EDX analysis and the XRD analysis, it is seen that the gel matrix consists of calcium silicate hydrate gel as well as a hybrid alumino–calcium-silicate gel (Figure 3.41f–j,where f,g,h,i and j represent 4% Na_2O,6% Na_2O,8% Na_2O,10% Na_2O and 12% Na_2O specimens respectively). The percentage of Ca ions shown in the EDX data decreases as leaching of calcium ions is inhibited by the high pH environment. As the alkali content is increased to 12%, there is deterioration in the gel matrix leading to cracking which is visible in the SEM image (Figure 3.53 e). As there are excess OH^- ions which affect the polymerization reaction in a negative manner, the calcium silicate hydrate is also not able to form. Because of the abundant OH^- ions, dissolution occurs very quickly and flash setting takes place and this impedes the polymerization reaction to a great extent resulting in the formation of cracks in the gel matrix.

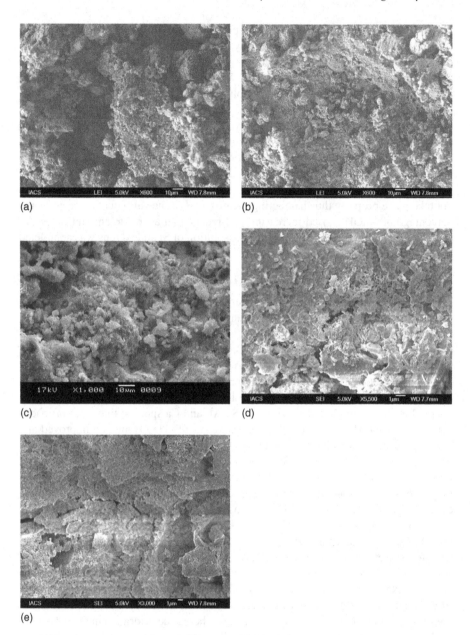

FIGURE 3.53 (a–e) SEM images for fly ash–slag AACs with different alkali *contents* (*%Na₂O*).

(*Continued*)

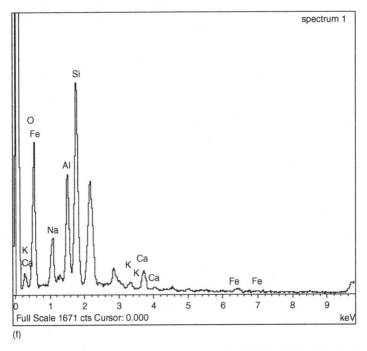

(f)

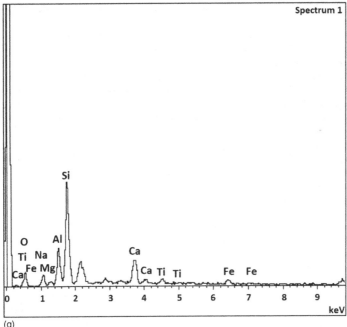

(g)

FIGURE 3.53 (CONTINUED)   (f–g) SEM images for fly ash–slag AACs with different alkali *contents* (%*Na₂O*).

(Continued)

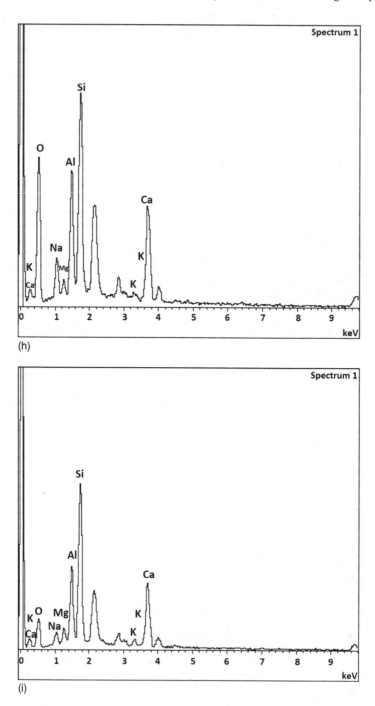

FIGURE 3.53 (CONTINUED)   (h–i) SEM images for fly ash–slag AACs with different alkali *contents* (%*Na₂O*).

(Continued)

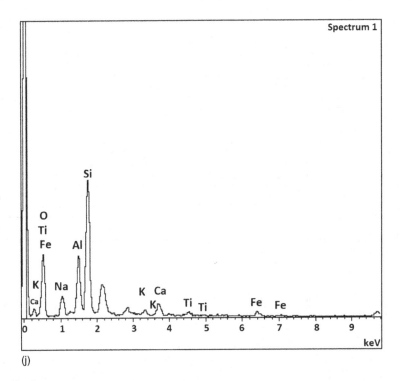

(j)

FIGURE 3.53 (CONTINUED)   (j) SEM images for fly ash–slag AACs with different alkali contents (*%Na2O*).

3.6.3.2.5 FTIR Analysis

On analyzing the FTIR data, it was seen from Figure 3.54 that there was a broad hump in the range of 3300 cm^{-1}–3700 cm^{-1} of all the samples which can be attributed to the OH^{-} group, the hump was wider in case of the A2(6% Na$_2$O), A3(8% Na$_2$O), and A4(10% Na$_2$O) samples. There was also a sharp peak intensity at 3704–3709 cm^{-1} which is due to the stretching vibration of the OH group. The hump noticed from 977 to 1134 cm^{-1} for A2 and A3 mix due to the asymmetric stretching vibration of Si-O-Si evolves into a sharp peak for A1(4% Na$_2$O) and A5(12% Na$_2$O) which signifies the incomplete polymerization process due to both excess and inadequate OH^{-} groups, respectively. The peaks at 470–480 cm^{-1}denoting bending Si-O-Si group vibration seem to shift to a lower wavenumber for A1 and A5 samples. Peaks due to C-S-H gel can be seen at 880–900 cm^{-1} which is a comparatively lower wavenumber as the C-S-H gel is normally seen at 900–950 cm^{-1} for alkali-activated slag composites. This is because as the samples are thermally cured, the aluminosilicate gel is functioning as the major gel network with C-S-H functioning as the minor one. The transmittance data in general point to the fact that the absorption values of the various functional groups have increased with the increase in sodium hydroxide

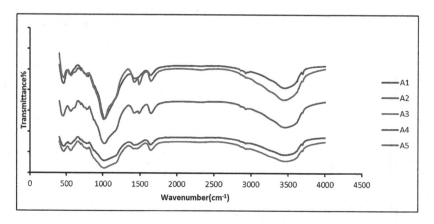

FIGURE 3.54 Effect of variation of alkali content (%Na$_2$O) on FTIR spectra of fly ash–slag-based AAC.

concentration up to the optimum level of 8% thus signifying a higher order of polymerization taking place in the A2 and A3 sample.

3.6.3.2.6 Zeta Potential and Electrical Conductivity

Zeta potential is an important factor governing electrostatic interactions in particle dispersions and as such, it plays a key role in understanding the stability of colloidal dispersions. It is used in formulating optimal solutions of suspensions and emulsions and can also be used to build a long-term stability prediction model [, 124,]. It has been known to have an effect on the microstructure of geopolymer specimens [125]. Analytical techniques such as XRD and SEM and its relationship with hardened and fresh properties of concrete have been conducted by the authors as well as other researchers [126]. In the current study, a relationship has been tried to establish between zeta potential and other mechanical properties of the test specimens. The zeta potential values of the fly ash and slag raw materials (before adding activator) were −9.44 mV and −13.9 mV, respectively (Figure 3.55). The negative zeta potential signified that the silicates (−O–SiO$_2$−) and aluminates (−O–Al–O−) were present abundantly near the surface. The negative zeta potential of the mixes A1, A2, A3, A4, and A5 are seen to increase with addition of sodium hydroxide. As the leaching rate of silicates and aluminates depends upon the rate of dissolution which increases with the addition of sodium hydroxide solution, it can be said that the negative zeta potential leads to faster dissolution. This aligns with fact that A5 having the highest negative zeta potential values had experienced flash setting which can be due to the greater rate of dissolution. The XRD results also point to the fact that the gel formation increased with higher negative zeta potential values as the peaks of aluminosilicate and calcium aluminosilicate had higher peaks. The higher zeta potential of the A5 mix could also be a result of the additional OH$^-$ ions of the mix which had negative impact on the structure of the AACs.

The electrical conductivity of the specimens is seen to increase with the increase in NaOH concentration. The conductivity recorded for the A1 specimen with 4% NaOH is the lowest at 0.0569 mS/cm, whereas the value increases significantly for

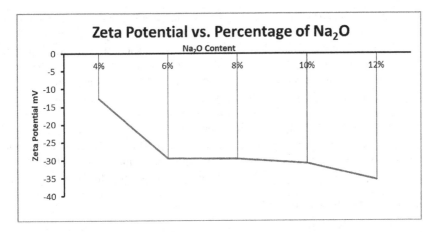

FIGURE 3.55 Effect of variation of alkali content (%Na$_2$O) on zeta potential of fly ash–slag-based AACs.

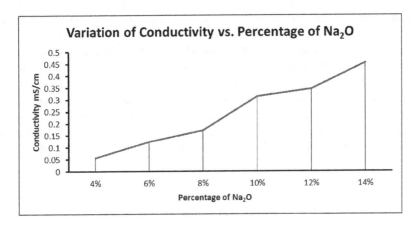

FIGURE 3.56 Effect of variation of alkali content *(%Na$_2$O)* on electrical conductivity of fly ash–slag AACs.

the A2 specimen at 0.123 mS/cm (Figure 3.56). The conductivity for the A3, A4, and A5 specimens continues to increase with recorded values of 0.172 mS/cm, 0.314 mS cm, and 0.345 mS/cm, respectively. The increase of conductivity of the specimens can be attributed to the formation of a denser microstructure with the increase in sodium hydroxide as visible in the SEM images. The increase in ions as well as ionic mobility due to the greater leaching of Si$^+$ and Al$^+$ ions may also influence the conductivity.

3.6.3.2.7 Summary

1. Experimental results indicate the co-existence of calcium silicate gel and alumino silicate gel.
2. The addition of sodium hydroxide up to a concentration of 8% results in the formation of an extremely dense and more compact microstructure and

provided a higher value of compressive strength, bulk density, apparent porosity, and water absorption as well as UPV values. This may be attributed to the increase of rate of dissolution with the increase in sodium hydroxide content, which leads to the leaching of Si^+, Al^+, and Ca^+ ions. However, excess sodium hydroxide (more than 8%) leads to flash setting and does not allow one to have a proper workable mix. It also impedes the formation of a calcium silicate hydrate gel and as a result, lower strength was observed.

3. The zeta potential value as well as electrical conductivity values increases with the increase in sodium hydroxide concentration. This can be attributed to the greater ionic mobility generated in the specimens due to the higher rate of leaching of Si^+ and Al^+ ions due to the increase of sodium hydroxide content.

3.6.3.3 Effect of Silica Content

3.6.3.3.1 Compressive Strength

The mix B1 with 4% SiO_2 registered the lowest compressive strength of 16 kN/m². The mix B2 with 6% SiO_2 recorded a growth of 86.85% over B1 (Figure 3.57). The mix B3 with 8% SiO_2 showed a growth of 19.31% over B2 but B4 (10% SiO_2) recorded a lesser growth of only 7.33% over B3 .The rate of strength increased by 1.39% from B4 to B5 (12% SiO_2),which recorded a strength of 39.47 kN/m². However, there is a drop in strength for mix B6 (14% SiO_2) which recorded a compressive strength of 24.4 kN/m². The increase of soluble silica in the polymeric system tends to increase the rate of polymerization specifically the process of condensation [127]. As the amount of soluble silica is increased, the polymerization process is accelerated to some extent. Furthermore, the presence of slag helps in contributing Ca^{2+} ions which reacts with anions present in the sodium silicate solution to form calcium silicate gel which helps strengthen the gel matrix [128].

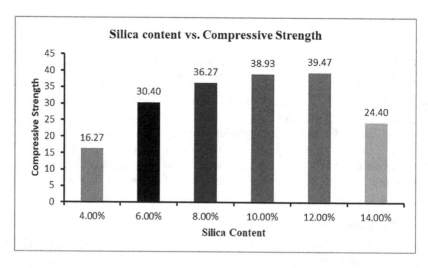

FIGURE 3.57   Effect of variation of silica content (%SiO_2) on compressive strength of fly ash–slag-based AACs.

At the beginning of the reaction, the vitreous component of the source material starts to precipitate forming a reaction product with the Si/Al ratio near to one [129,130,131]. It has been seen in previous studies that with the gradual progression of the reaction the aluminum sources in the system get depleted and the silica content starts to rise thus giving rise to a higher Si/Al ratio in the final geopolymer product [132]. The presence of soluble silicate in the geopolymeric system is extremely essential for initiation of formation of oligomers which consist of monomers, dimers, trimers, and tetramers. The increase in silica content in the mix proportion ensures the formation of oligomers from the beginning of the geopolymeric reaction which fosters polycondensation at an early stage [133]. The abovementioned phenomenon causes an increase in the mechanical properties of the AACs but the percentage up to which the silica content can be added has a threshold value which is 12% noticed in the present study. Increasing the silica content beyond the threshold value has a negative impact on the compressive strength of the AAC specimens. One of the reasons behind this is the great increase in viscosity of the fresh AACs which makes it extremely difficult to cast them into molds and due to this the hardened AAC cubes have various formation defects which lead to a fall in strength .Another important parameter guiding polymerization is the rate of dissolution which increases with addition of sodium hydroxide content. In the present study, the Na_2O content is fixed at 8% which promotes a speedy dissolution. The stages of polymerization can take place discretely as well as simultaneously, and the presence of increased silica content with high alkali content will give rise to dissolution and polycondensation simultaneously which will be highly beneficial to the development of strength characteristics of AACs.

3.6.3.3.2 Bulk Density, Water Absorption, Apparent Porosity, and UPV

The bulk density for the B1 mix with 4% SiO_2 is seen to have the lowest bulk density value at 1744.22 kg/m^3 (Figure 3.58). The value increases to 1784.77 kg/m^3 for B2 mix with 6% SiO_2. The bulk density value for B3 mix with 8% SiO_2 is a bit higher at 1842.262 kg/m^3. It continues to increase for B4 (10% SiO_2) mix and B5 (12% SiO_2) mix with 1871.92 kg/m^3 and 1878.33 kg/m^3, but it registers a drop for the B6(14% SiO_2) mix with a value of 1748.63 kg/m^3. The rate of polymerization increases with the rate of dissolvable silica and aids in gel formation. The addition of silica here serves a dual purpose of forming sialate and polysialate atoms, helping to form a 3-D aluminosilicate gel and also caters to the formation of a calcium silicate hydrate gel. This calcium silicate gel is formed due to the reaction of calcium ions of the slag coming in contact with the silicate species. Thus, the formation of both aluminosilicate and calcium silicate-based gel contributes to better bulk density and also influences the other mechanical properties as well. The apparent porosity decreases with the increase of silica content up to the threshold level of 12%. The water absorption percentages also tend to decrease with the addition of silicate species. B1 samples showed water absorption values of 5.28% highest whereas for remaining specimens B2, B3, B4, B5, and B6, recording values of 4.18%, 2.48%, 1.32%, 1.1%, and 4.82%, respectively (Figure 3.59). The results align with the trend of bulk density values. Water absorption values also follow the same trend as the bulk density and apparent porosity values (Figure 3.60). The main reason behind the observed trends of the

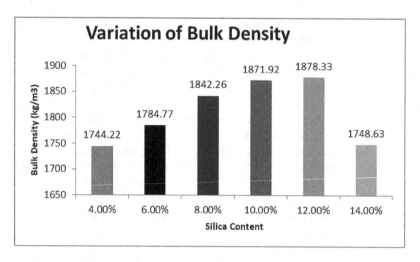

FIGURE 3.58 Effect of variation of silica content (%SiO₂) on bulk density of fly ash–slag-based AACs.

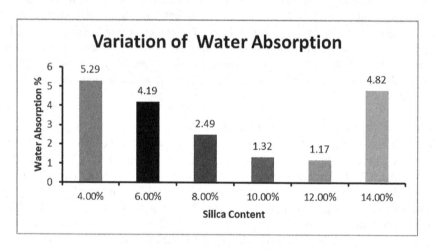

FIGURE 3.59 Effect of variation of silica content (%SiO₂) on water absorption of fly ash–slag-based AACs.

different mechanical properties is lessening of voids in the structure of the specimens due to addition of silica. The changes in the abovementioned mechanical properties can be corroborated with UPV values as well. The UPV values of the B1, B2, B3, B4, B5, and B6 mixes are 3.78 km/s, 4.12 km/s, 4.16 km/s, 4.25 km/s, 4.58 km/s, and 3.96 km/s, respectively (Figure 3.61). It can be said from the above-discussed results that the addition of silica up to an optimum level results in the formation of a closely spaced, dense and cohesive microstructure. This dense cohesive microstructure also has its roots in the usage of both calcium-based and alumina-silicate-based source materials being used in the present study. The increase of SiO_2/Na_2O ratio sets into place a gradual shift of the chemical system to a 3-D polymeric framework which

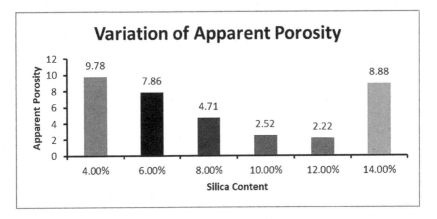

FIGURE 3.60 Effect of variation of silica ($\%SiO_2$) on apparent porosity of fly ash–slag-based AACs.

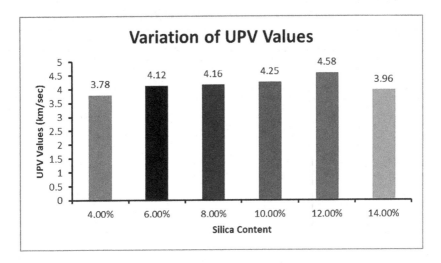

FIGURE 3.61 Effect of variation of silica ($\%SiO_2$) on the UPV value of fly ash–slag-based AACs.

enhances the mechanical properties of AACs [134]. This can further be explained by the XRD analysis and FESEM images in the subsequent sections.

3.6.3.3.3 Mineralogical Investigation by XRD Analysis

XRD data of the various mix specimens (Figure 3.62a–f) suggest that no major phase change occurs with the variation of silica content .The main reaction products were quartz and mullite with sillimanite and mullite constituting the minor products. The same set of reaction products was found in all the specimens. The intensity of quartz though began to lessen with the increase in silica content; this is possibly due to the greater rate of polycondensation with addition of silica content. There is also the presence of calcium alumino silicate gel as well in all the specimens, which is due to the presence of slag.

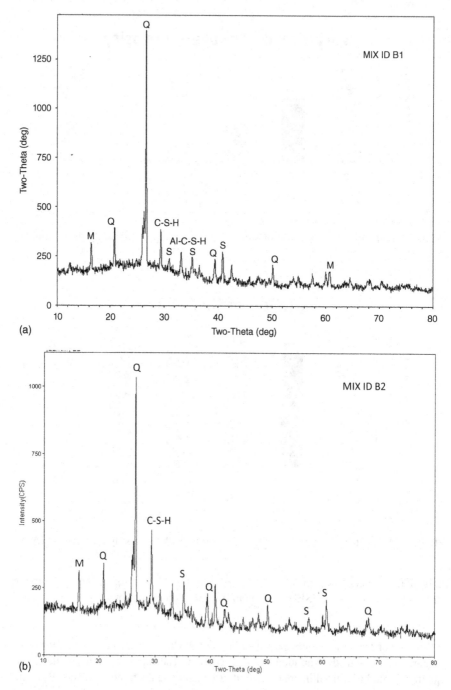

FIGURE 3.62   (a–b) Effect of variation of silica content (%SiO₂) on XRD graphs of fly ash–slag-based AACs.

(*Continued*)

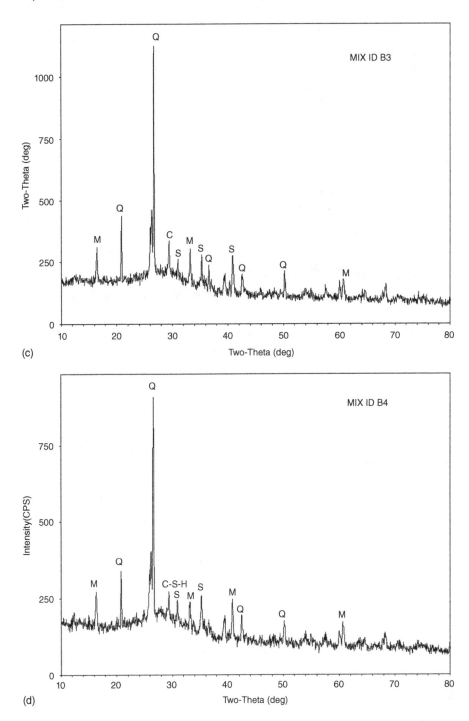

FIGURE 3.62 (CONTINUED) (c–d) Effect of variation of silica content (%SiO2) on XRD graphs of fly ash–slag-based AACs.

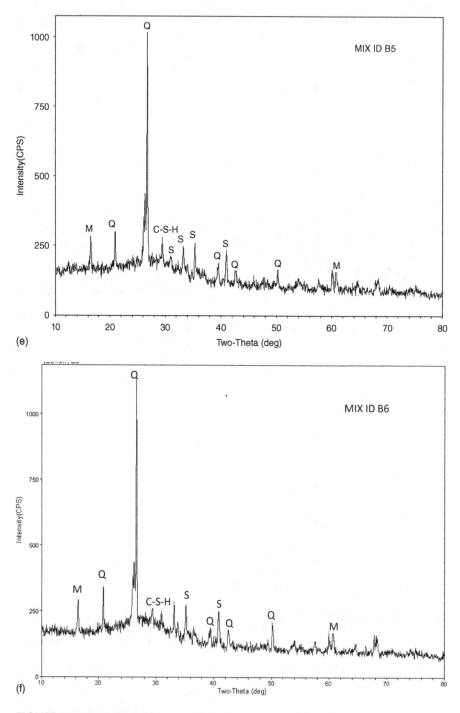

FIGURE 3.62 (CONTINUED)   (e -f) Effect of variation of silica content (%SiO2) on XRD graphs of fly ash–slag-based AACs..

3.6.3.3.4 Microstructure Study by SEM/EDX

It was seen in the SEM images and corresponding EDX data (Figures 3.63a–f and Figures 3.63g–l,where a,b,c,d,e and f and g,h,i,j,k and l represent 4% SiO_2,6% SiO_2,8% SiO_2,10% SiO_2, 12% SiO_2,14% SiO_2 specimens respectively). that the addition of silicate has an effect on the intensity of polymerization and also its subsequent reaction products. This can also be clearly seen in the SEM images. SEM images of the B1 mix show the presence of spherical unreacted fly ash particles surrounded by a disjointed gel matrix. In the B2 mix, a formation of a more consistent gel matrix was noticed. B3 mix displays a more cohesive and denser microstructure, and more products of precipitation were noticed which point to an increased rate of condensation .The products of condensation increases in the samples B4 and B5. But in the sample B6 containing 14% SiO_2, the gel matrix is not properly formed in presence of multiple unreacted fly ash particles and prominent voids. Another point worth mentioning is that the amount of aluminum content in the specimens starts to increase with the silicate. This is due to the leaching of Al+ ions by the respective activator solution. The homogeneity of the microstructure continues to improve with the addition of silicate ranging from 4% to 12%. Si-O-Si bonds are produced in greater quantity with the increase of Si/Al ratio [135] and as Si-O-Si linkages exhibit a stronger behavior than Si-O-Al and Al-O-Al bonds [136], an increase in the strength of the AACs with increasing silicate is expected but that is not the case for 14% silica content. This is specifically due to the generation of a very stiff mix which leads to improper compaction and subsequently leads to the formation of more pores and a weak polymeric matrix.

3.6.3.3.5 TGA/DTA

On analyzing the TGA data of the various mix specimens, it is seen that the mass loss percentage decreases with the increase of silica content (Figure 3.64). As seen in the SEM images, the specimens with a greater silica content reported a denser and more cohesive microstructure. As the microstructure becomes denser, the pore volume decreases as well and with it the amount of water stored in pores also lessens, and thus, the samples with a more silica percentage report a lesser mass loss percentage value.

3.6.3.3.6 Summary

1. The increase in sodium silicate percentage accelerates the rate of polycondensation which starts happening at an earlier stage of polymerization almost simultaneously with dissolution.
2. The presence of soluble silica is extremely essential for the formation of oligomers, for example, monomers, dimmers, and trimers.
3. The increased rate of polycondensation gives rise to a denser and more cohesive microstructure. It also results in an increase of values of mechanical and engineering properties of the AAC specimens, for example, compressive strength, bulk density, water absorption, and apparent porosity.
4. Major phase changes were not noticeable in the X-ray diffractograms of the mix specimens.
5. TGA/DTA pointed to the fact that the presence of a higher percentage of sodium silicate leads to the presence of less freely available water thus causing a lower mass loss of test samples with a higher silicate content with the increase of temperature level.

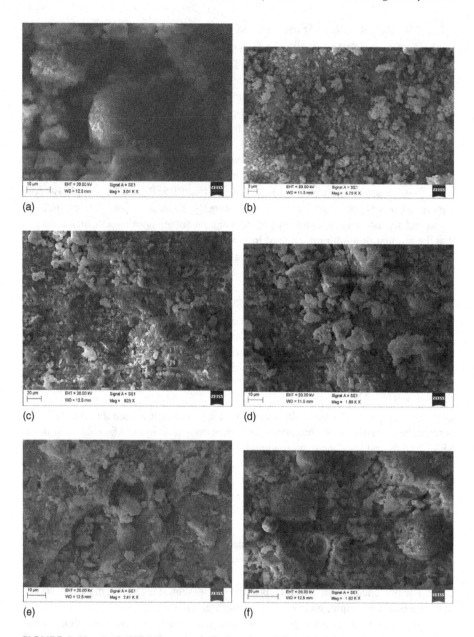

FIGURE 3.63 (a–f) SEM images of alkali-activated fly ash–slag AACs with variation of silica content. (%SiO$_2$)

(*Continued*)

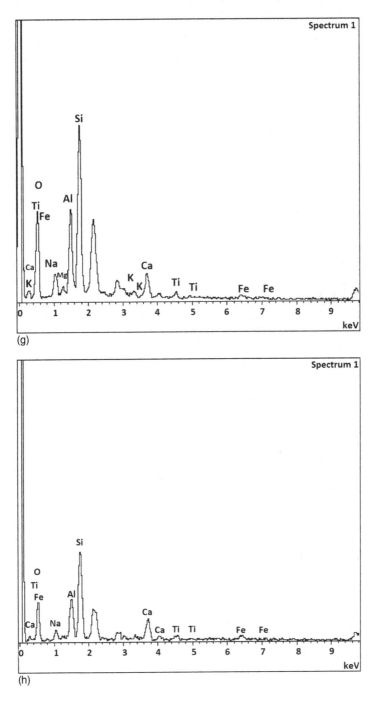

FIGURE 3.63 (CONTINUED) (g–h) EDX data for alkali-activated fly ash–slag AACs with variation of silica content. (%SiO$_2$)

(Continued)

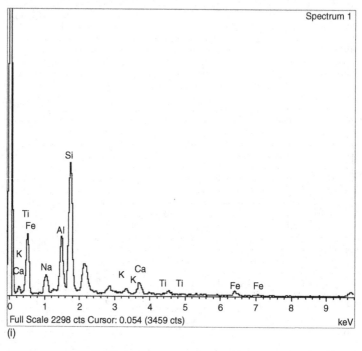

(i)

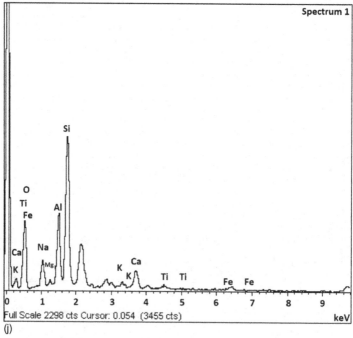

(j)

FIGURE 3.63 (CONTINUED) (i -j) SEM images of alkali-activated fly ash–slag AACs with variation of silica content. (%SiO$_2$).

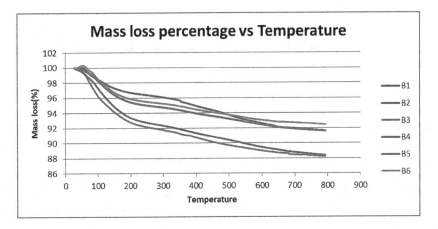

FIGURE 3.64   Effect of variation of silica content (SiO_2) on mass loss of fly ash–slag-based AACs.

3.6.3.4 Effect of Water Content

3.6.3.4.1 Compressive Strength

The water to binder ratio has been changed from 0.35 to 0.47, keeping SiO_2 and Na_2O constant at 8%. Water content was expressed in terms of w/b ratio by weight.

Figure 3.65 indicates that the compressive strength increased as the w/b ratio was increased from 0.35 to 0.38. However, with further increase in the w/b ratio, reduction in compressive strength was observed. In the polymerization reaction of alkali-activated materials, water does not participate in the reaction directly but acts as a transportation medium, thus increasing the space in between gel particles. Furthermore, when the AACs are subjected to high curing temperature, the water evaporates, leaving behind voids; thus strength is decreased. For proper casting, placing, and compaction, an optimum w/b ratio of 0.38 is needed. The w/b ratio lesser

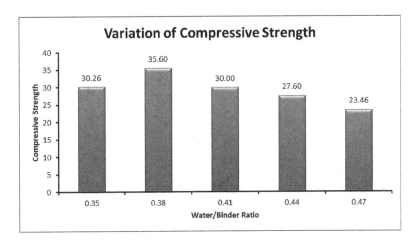

FIGURE 3.65   Effect of the w/b ratio on compressive strength of fly ash–slag-based AACs.

than 0.38 leads to a decrease in the mechanical properties of the AACs. Highest and lowest compressive strength was recorded for the w/b ratio of 0.38 and 0.47, respectively. The w/b ratio has to be judiciously chosen to suit desired workability of the mix as well to achieve higher strength.

3.6.3.4.2 Bulk Density, Apparent Porosity, Water Absorption, and UPV Value

The bulk density, apparent porosity, and water absorption values increase when the w/b ratio is increased from 0.35 to 0.38 (Figures 3.66, 3.67, 3.68, and 3.69). The mix was not workable at a w/b ratio 0.35, leading to improper casting and compaction resulting in the formation of more voids. A higher percentage of voids also causes a decrease in the bulk density, apparent porosity, and water absorption. However beyond w/b ratio 0.35, the excess water hinders the polymerization process thus negatively affecting the engineering properties. The maximum and minimum values for

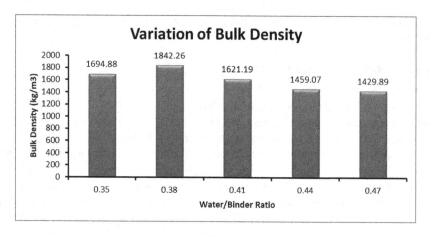

FIGURE 3.66   Effect of the w/b ratio on bulk density of fly ash–slag-based AACs.

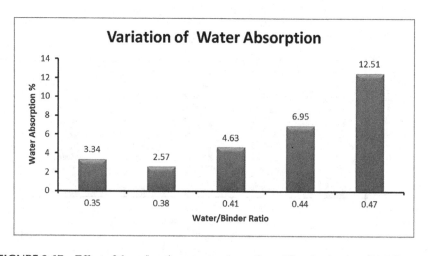

FIGURE 3.67   Effect of the w/b ratio on water absorption of fly ash–slag-based AACs.

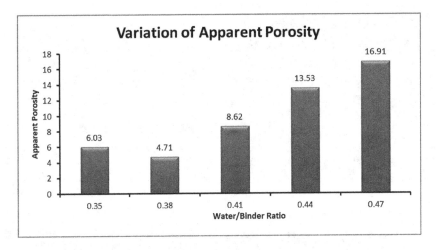

FIGURE 3.68 Effect of the w/b ratio on apparent porosity of fly ash–slag-based AACs.

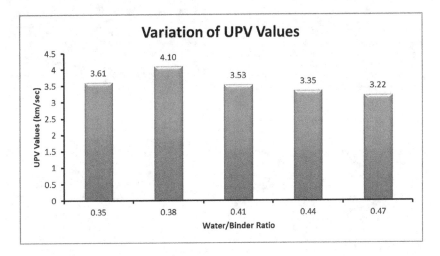

FIGURE 3.69 Effect of the w/b ratio on the UPV value of fly ash–slag-based AACs.

bulk density, apparent porosity, and water absorption were 1842.26 kg/cum, 2.57%, and 4.71% and 1429.59 kg/cum, 12.51% and 25.17% for 0.38 and 0.47 w/b ratio, respectively. These values are also supported by the UPV values which also report maximum velocity for the sample with a w/b ratio of 0.38.

3.6.3.4.3 SEM/EDX Analysis

On observation of SEM images and EDX analysis shown in Figure 3.70a–f and Figure 3.70g–h,(where a,b,c,d,e and f,g,h,i,j,k and l represents specimens having w/b ratio of 0.35,0.38,0.41,0.44,0.47 and 0.50 respectively), it is seen that the increase of w/b ratio beyond 0.38 leads to a decrease in the rate of polymerization which leads to the formation of voids and cracks. At a w/b ratio of 0.47, it was seen that there were unreacted fly ash particles present in the gel structure.

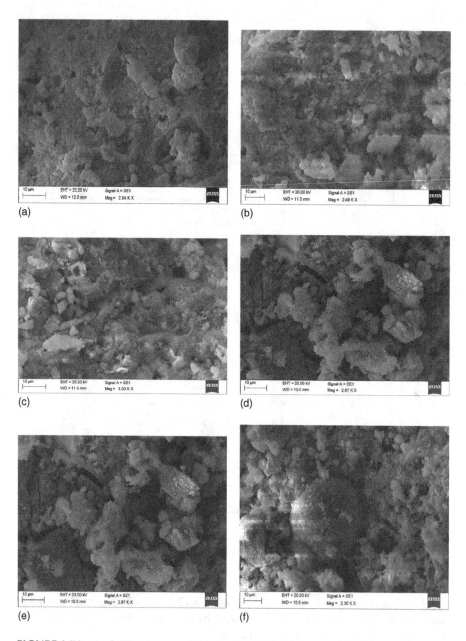

FIGURE 3.70   (a–f) SEM images of fly ash–slag AAC specimens with different w/b ratios.

(*Continued*)

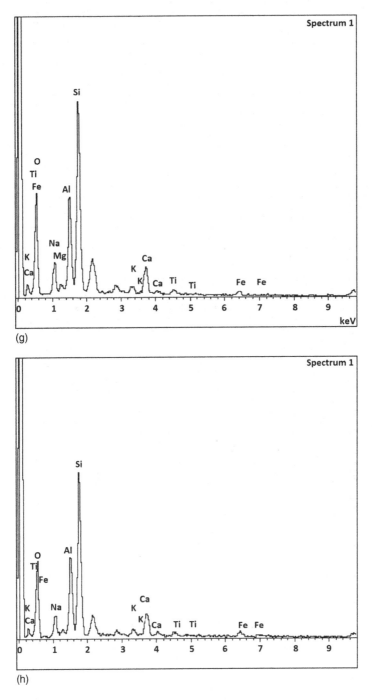

FIGURE 3.70 (CONTINUED) (g–h) EDX data for fly ash–slag AAC specimens with different w/b ratios.

(*Continued*)

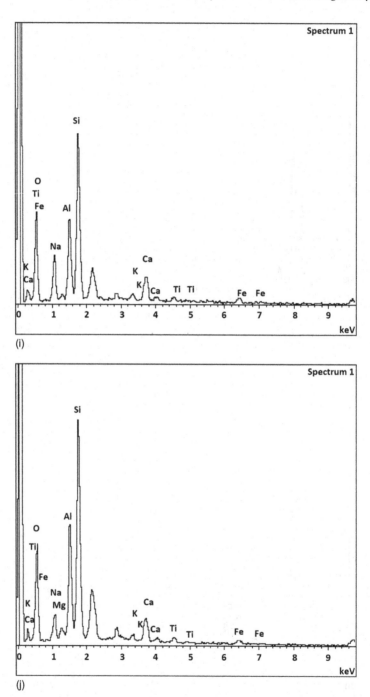

FIGURE 3.70 (CONTINUED)   (i–j) SEM images of fly ash–slag AAC specimens with different w/b ratios.

(*Continued*)

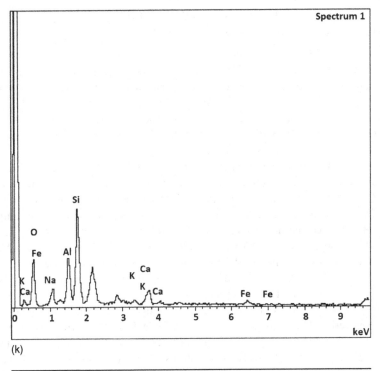

(k)

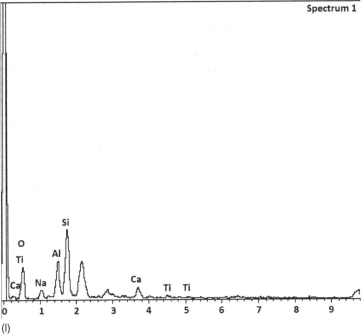

(l)

FIGURE 3.70 (CONTINUED) (k–l) SEM images of fly ash–slag AAC specimens with different w/b ratios.

3.6.3.5 Effect of Curing (Thermal and Water Curing)

3.6.3.5.1 Effect of Thermal Curing

3.6.3.5.1.1 Compressive Strength

The effect of curing temperature (oven curing) on compressive strength was investigated in this test series. The oven curing was performed at 60°C and 85°C .The specimens were kept in the oven for a duration of 48 h and then left in air for 7 days and 28 days. The compressive strengths of the specimens are given in Table 3.18. It may be noted here for better understanding that M 90-10-85 indicates that fly ash 90%, slag 10%, and oven curing temperature of 85°C.

Thermal curing has been carried out on the AAC samples for the dual cause of strength gain and achievement of the polymeric gel network. Temperature was fixed at 60°C and 85°C. From the point view of strength, the samples cured at a temperature of 60°C had a higher strength gain with increasing slag content. The sample M 50-50-60 recorded the highest strength of 52.09 MPa [Figure 3.71 and 3.72]. But when the temperature was 85°C, the sample M 70-30-85 recorded the highest strength of 40.18 Mpa. The M-50-50-85 sample reports a slightly lesser compressive strength of 39.122 MPa. At a temperature of 85°C, microcracks were observed in the samples M-50-50-85 and M-40-60-85.

The effect of duration of thermal curing on compressive strength was investigated in this test series. The results are tabulated in Table 3.19. The curing duration varied from 24 to 72 h. Figures 3.73, 3.74, 3.75, and 3.76 present the compressive strength of AAC paste specimens for different durations of oven heat curing at 85°C. It can be noticed that the compressive strength of the test specimens was significantly affected by the duration of thermal curing and slag content. It can be observed from Figure 3.73 that the compressive strength increases up to a curing temperature of 85°C for slag content up to 30%.When the slag content is more than 30%, the strength starts to decrease. It had been seen in the earlier discussion that 85°C was not conducive to

---

TABLE 3.18

Effect of Curing Temperature on Compressive Strength

Mix ID	%Na$_2$O in Activator	%SiO$_2$ in Activator	w/b Ratio	Oven Curing Temperature	7 Day (MPa)	28 Day (MPa)
M 90-10-85	8	8.0	0.35	85°C	24.55	28.23
M 85-15-85	8	8.0	0.35	85°C	28.47	34.164
M 70-30-85	8	8.0	0.35	85°C	32.67	40.18
M 50-50-85	8	8.0	0.35	85°C	31.55	39.122
M-40-60-85	8	8.0	0.35	85°C	29.55	37.82
M-90-10-60	8	8.0	0.35	60°C	27.01	31.87
M-85-15-60	8	8.0	0.35	60°C	33.59	41.32
M-70-30-60	8	8.0	0.35	60°C	40.18	51.43
M-50-50-60	8	8.0	0.35	60°C	40.38	52.09
M-40-60-60	8	8.0	0.35	60°C	38.42	49.94

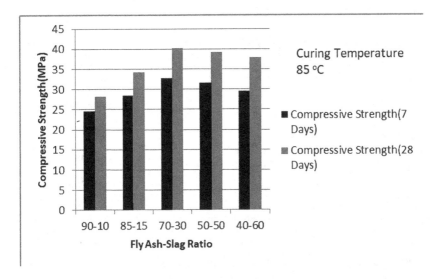

FIGURE 3.71   Effect of thermal curing at a temperature on compressive strength.

specimens with a high slag content. It was seen that as the curing duration is increased, the compressive strength continues to decrease. The maximum compressive strength of 36.65 MPa at a curing temperature of 85°C was noticed for sample 70-30-85 cured for 72 h. The specimens subjected to a curing temperature of 60°C reported a different trend compared to the specimens cured at 85°C. At a curing temperature of 60°C, the specimens reported a continuous increase in compressive strength with the

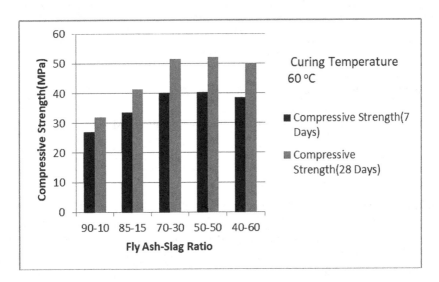

FIGURE 3.72   Effect of thermal curing at a temperature on compressive strength.

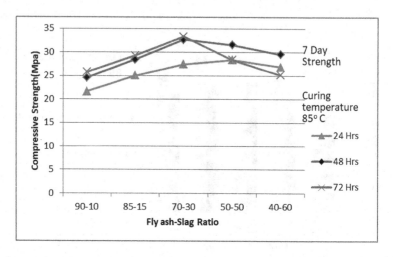

FIGURE 3.73   Effect of duration of thermal curing at 85°C.

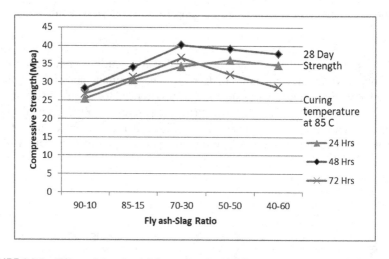

FIGURE 3.74   Effect of duration of thermal curing 85°C.

increase in slag content up to 50% for all curing durations. Mix 50-50-60 reported the highest compressive strength of 57.82 MPa.

3.6.3.5.2 Effect of Water Curing

3.6.3.5.2.1 Compressive Strength

Addition of slag was noticed to have a positive effect with regard to compressive strength as can be seen in Figure 3.77. The increase of compressive strength can be credited to the more reactive character of the slag particles which leads to the formation of a hybrid N-C-A-S-H gel [108–109]. The bar chart indicates that the mix M-40-60 reports the highest compressive strength of 42.4 MPa and 58.512 MPa at 7 and 28 days, respectively. Correspondingly, the mix 90-10 reports the lowest

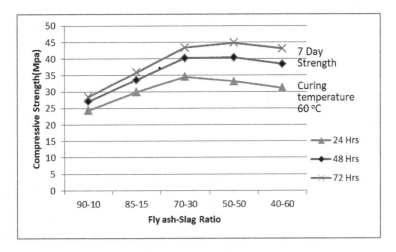

FIGURE 3.75 Effect of duration of thermal curing at 60°C.

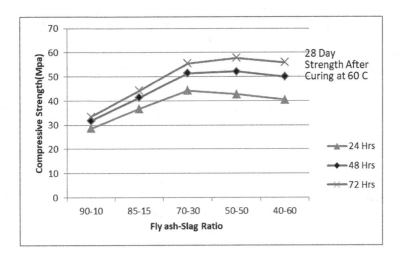

FIGURE 3.76 Effect of duration of thermal curing at 60°C.

compressive strength of 21.6 and 25.92 at 7 and 28 days, respectively. Similar results have been reported by Kumar et al. [23] who also reported an increase in compressive strength with addition of slag with fly ash. Yip et al. [109] reported that the coexistence of both C-S-H and geopolymeric gel could be a reason behind the increase of compressive strength of an AAC paste.

3.6.3.5.2.2 Bulk Density, Water Absorption, Apparent Porosity, and UPV
It can be seen from Figure 3.78 that the bulk density increases with the increase of percentage of slag content. Mix M-90-10 recorded the lowest bulk density values of 1565.66 kg/cum and 1643 kg/cum at 7 and 28 days, respectively, the highest

TABLE 3.19
Effect of Curing Duration on Compressive Strength

Mix ID	%Na_2O in Activator	%SiO_2 in Activator	w/b Ratio	Oven Curing Temp	Duration	7 Day (MPa)	28 Day (MPa)
M 90-10-85	8	8	0.38	85°C	24 h	21.604	25.49
M 85-15-85	8	8	0.38	85°C	24 h	25.05	30.56
M 70-30-85	8	8	0.38	85°C	24 h	27.44	34.303
M 50-50-85	8	8	0.38	85°C	24 h	28.395	36.06
M-40-60-85	8	8	0.38	85°C	24 h	26.89	34.688
M-90-10-85	8	8	0.38	85°C	48 h	24.55	28.23
M-85-15-85	8	8	0.38	85°C	48 h	28.47	34.164
M-70-30-85	8	8	0.38	85°C	48 h	32.67	40.18
M-50-50-85	8	8	0.38	85°C	48 h	31.55	39.122
M-40-60-85	8	8	0.38	85°C	48 h	29.55	37.824
M-90-10-85	8	8	0.38	85°C	72 h	25.78	26.808
M-85-15-85	8	8	0.38	85°C	72 h	29.32	31.37
M-70-30-85	8	8	0.38	85°C	72 h	33.32	36.66
M-50-50-85	8	8	0.38	85°C	72 h	28.395	32.08
M-40-60-85	8	8	0.38	85°C	72 h	25.11	28.63
M 90-10-60	8	8	0.38	60°C	24 h	24.309	28.68
M 85-15-60	8	8	0.38	60°C	24 h	29.89	36.77
M 70-30-60	8	8	0.38	60°C	24 h	34.55	44.23
M 50-50-60	8	8	0.38	60°C	24 h	33.116	42.71
M-40-60-60	8	8	0.38	60°C	24 h	31.12	40.46
M-90-10-60	8	8	0.38	60°C	48 h	27.01	31.87
M-85-15-60	8	8	0.38	60°C	48 h	33.59	41.32
M-70-30-60	8	8	0.38	60°C	48 h	40.18	51.43
M-50-50-60	8	8	0.38	60°C	48 h	40.38	52.09
M-40-60-60	8	8	0.38	60°C	48 h	38.42	49.94
M-90-10-60	8	8	0.38	60°C	72 h	28.36	33.46
M-85-15-60	8	8	0.38	60°C	72 h	35.94	44.21
M-70-30-60	8	8	0.38	60°C	72 h	43.39	55.54
M-50-50-60	8	8	0.38	60°C	72 h	44.82	57.82
M-40-60-60	8	8	0.38	60°C	72 h	43.03	55.93

being recorded by the mix M-40-60 with bulk density value of 1729.54 kg/cum and 1816.02 kg/cum for 7 and 28 days, respectively. The same trend akin to the bulk density values can be noticed in the water absorption, apparent porosity, and UPV values shown in Figures 3.79, 3.80, and 3.81. All these properties were found to improve with addition of slag. The UPV values of the samples at 14 days were 4.09 km/s, 4.54 km/s, 4.62 km/s, 4.61 km/s, and 4.89 km/s for mix M-90-10, M-85-15, M-70-30, M-50-50, and M-40-60, respectively. These values point to the

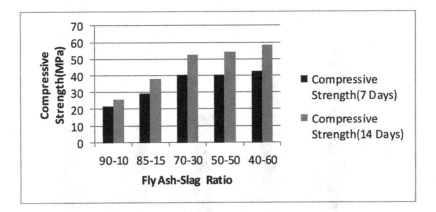

FIGURE 3.77 Effect of variation of slag on compressive strength of fly ash–slag-based AACs.

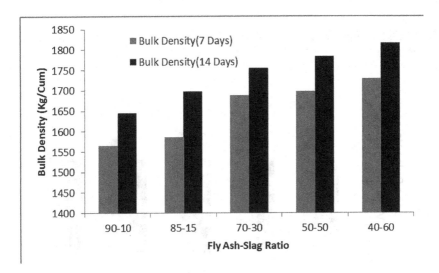

FIGURE 3.78 Effect of variation of slag on bulk density of fly ash–slag-based AACs. (water cured)

increase in density of the AAC paste. When the paste becomes denser, it also lessens the water absorption and apparent porosity values of the samples. A major factor contributing to this phenomenon is the choice of curing method which is water curing in this case. As fly ash-based AACs generally require heat curing for gel solidification to take place, they do not fully participate in the reaction under water curing. However, slag particles have calcium as their main constituent which makes them more reactive in the presence of water and in general, it increases the reactivity of the paste as a whole. The changes in the gel network can be further

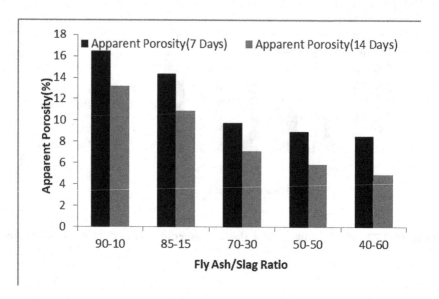

FIGURE 3.79 Effect of variation of slag on apparent porosity of fly ash–slag-based AACs. (water cured)

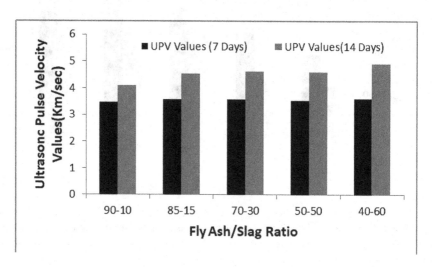

FIGURE 3.80 Effect of variation of slag on the UPV value of fly ash–slag-based AACs. (water cured)

explained with the XRD, FTIR, and FESEM/EDX analysis discussed in the following sections.

3.6.3.5.2.3 Mineralogical Investigation by XRD Analysis

XRD data presented in Figure 3.82a and b show peaks of SiO_2, mullite, sillimanite, and a phase which may be classified as sodium-alumino-calcium-silicate-hydrate

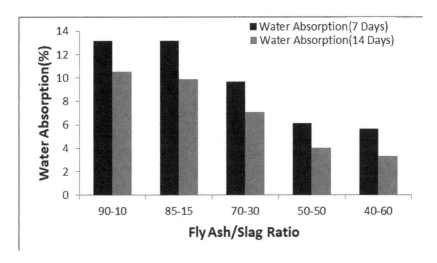

FIGURE 3.81 Effect of variation of slag on water absorption of fly ash–slag-based AACs. (water cured)

(N-Al-Ca-S-H) gel. The peak corresponding to the N-Al-Ca-S-H gel increases in intensity with the increase in the slag content which signifies that the greater amount of N-Al-Ca-S-H gel is formed with the increase in slag content .This gel contributes in making the paste more dense which in turn leads to low apparent porosity and water absorption values

3.6.3.5.2.4 Microstructure Study by SEM/EDX

In Figure 3.83a, it is noticed that the gel is severely disjointed in nature, and unreacted fly ash particles are visible. In Figure 3.83b, the gel structure is less disjointed in nature, and fewer spherical fly ash particles are visible. Gradually, with the addition of slag, the gel structure becomes more and more compact in nature and also unreacted fly ash particles become less. In Figure 3.83c, it is noticed that the spherical fly ash particle is covered in a sheath of reaction products but in the mix M-50-50 and M-40-60 (Figure 3.83d, e), it is observed that the amount of gel formation has increased significantly and unreacted fly ash particles are no longer visible. The gel that is visible is clearly sodium-alumino-calcium-silicate gel which can be deduced from the XRD results where the peaks of the C-A-S-H gel become sharper and more distinct with the increase of slag content. The SEM images pointed to the formation of a denser microstructure which is due to the formation of a cross-linked polymeric N-C-A-S-H gel. Increasing cross-linkage improves the mechanical properties of the composites. The nature of the binding gel formed depends upon the slag content in the mix. The increase of slag content leads to the inclusion of calcium ions into the polymeric gel network. It also lessens the formation of pores in the matrix, thus decreasing the apparent porosity and water absorption values of the composites.

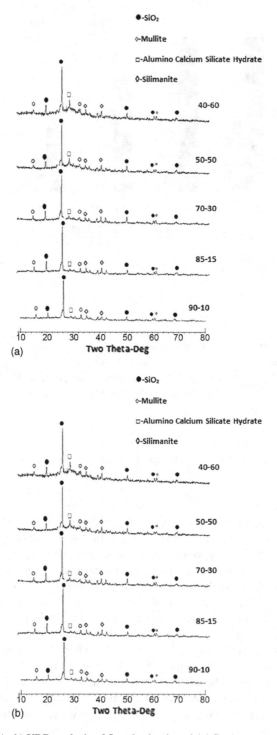

FIGURE 3.82 (a, b) XRD analysis of fly ash–slag-based AACs. (water cured)

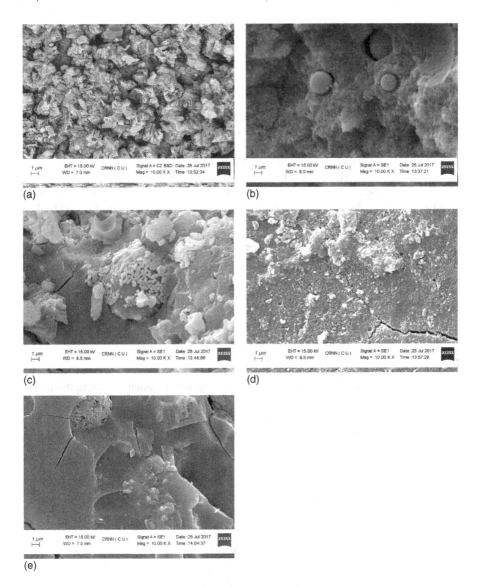

FIGURE 3.83 (a–e) SEM images for fly ash–slag-based AACs with different slag contents.

3.6.3.6 Effect of Fineness of Slag

In this series of tests, the effect of fineness of GGBS on properties of the AAC specimens was studied. GGBS was prepared by grinding and sieving into the following three groups according to their fineness. The mix proportion is given in Table 3.20.

Sieving of fineness type: Particle size: < 45, < 75, and <150 µm.

TABLE 3.20

Mix Proportions to Study the Effect of Fineness of GGBS on Compressive Strength of Blended AACs

MIX ID	Source Material Ratio (wt.% Fly Ash/wt.% Slag)	Sodium Silicate (Activator) Dose Relative to Binder Content (wt.%)	Sodium Hydroxide (Activator) Dose Relative to Binder Content (wt.%)	Water/ Binder Content	Fineness of Slag	Fineness of Fly Ash
SL150-FL150	70-30	8	8	0.38	<150 μm	<150 μm
SL75-FL150	70-30	8	8	0.38	75–150 μm	<150 μm
SL45-FL150	70-30	8	8	0.38	45–75 μm	<150 μm

3.6.3.6.1 Compressive Strength

The compressive strength of the specimens improved with the increase in fineness of slag particles as can be seen in Figure 3.84. The improvement in strength can be attributed to the increase in reactivity of the slag material with the increase in fineness. It can be seen that the highest compressive strength of 35 Mpa and 47 Mpa at 7 and 14 days, respectively, has been reported by the mix id SL45-FL150 containing

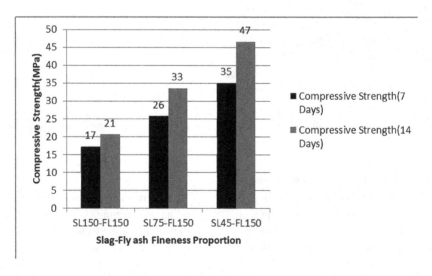

FIGURE 3.84 Effect of variation of slag on bulk density of fly ash–slag AACs.

slag particles of fineness between 45 and 75 micron. Correspondingly, the mix SL150-FL150 recorded the lowest compressive strength of 17 Mpa and 21 Mpa, respectively. Similar results have been reported by Qureshi et al [137] who also reported an increase in compressive strength of alkali-activated slag composites with the increase in fineness of slag particles. Further discussion about the other engineering properties is done in the following section.

3.6.3.6.2 Bulk Density, Water Absorption, Apparent Porosity, and UPV

It is observed from Figure 3.85 that on increasing the fineness of slag particles, there is an increase in the bulk density. SL-45-FL150 reported the highest bulk density values of 1688 kg/cum and 1877 kg/cum at 7 and 14 days, respectively, and the lowest being recorded by the mix SL150-SL150 with bulk density value of 1556 kg/cum and 1636 kg/cum for 7 and 28 days, respectively. The improvement in bulk density is in sync with the results of the water absorption and apparent porosity of the mix specimens. Both water absorption and apparent porosity were found to decrease with the increase in fineness of slag particles. The increase in bulk density combined with the decrease in water absorption and apparent porosity is further supported by the UPV values of the test specimens at 14 days. The highest UPV value was reported by the mix ID SL45-FL150, which was 3.78 km/s and 3.85 km/s for 7 and 14 days, respectively, (Refer Figure 3.87). The increase in fineness of slag leads to a greater surface area of the particles available for polymerization thus leading to increased gel formation, which in turn lessens the percentage of voids leading to a greater bulk density and decreased water absorption and apparent porosity as can be seen in Figures 3.86 and 3.88. One of the most important factors contributing to the increase in mechanical properties is mode of curing. As water

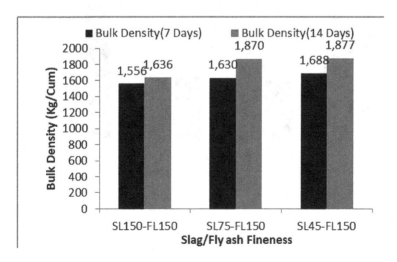

FIGURE 3.85 Effect of variation of fineness of slag on bulk density of fly ash–slag-based AACs.

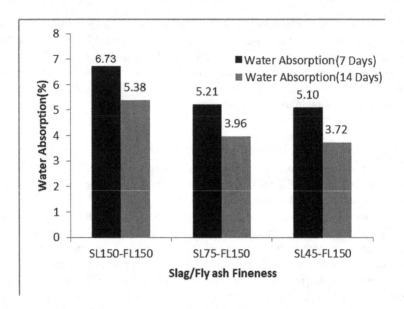

FIGURE 3.86 Effect of variation of fineness of slag on water absorption of fly ash–slag-based AACs.

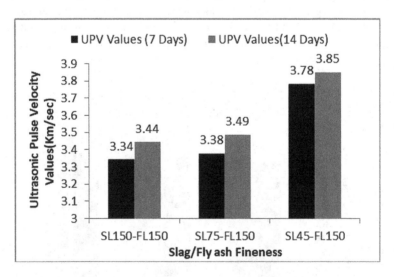

FIGURE 3.87 Effect of variation of fineness of slag on UPV values of fly ash–slag-based AACs.

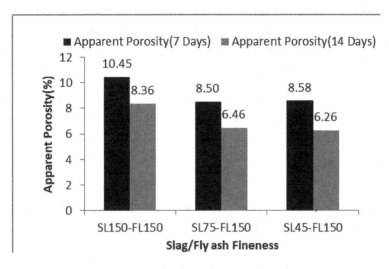

FIGURE 3.88 Effect of variation of fineness of slag on apparent porosity of fly ash–slag-based AACs.

curing has been used, the progress of the polymerization reaction depends more upon the reactivity of the slag particles rather than fly ash particles which would not have been the case if the thermal curing method was used. The reaction process and its resultant products can further be explained by SEM and XRD analysis in the following sections.

3.6.3.6.3 Mineralogical Investigation by XRD

XRD data for 7 and 14 days are quite similar, with both graphs recording peaks of SiO_2, mullite, and sillimanite and a phase which may be classified as Sodium-Alumino-Calcium-Silicate-Hydrate (N-A-C-S-H) gel. The sodium-alumino-calcium-silicate-hydrate peak increases in intensity with the increase in the fineness of slag, which signifies that there is an increase in gel formation with the increase in fineness of slag particles. This gel functions as the main polymeric binding gel and the increase in its peak corresponds to the improvement of the mechanical properties of the specimens (Figure 3.89).

3.6.3.6.4 Microstructure Study by SEM/EDX

In Figure 3.90c, it was noticed that there are large voids present in the gel structure. However with the increase in fineness of the slag particles, it is noticed that the gel matrix becomes more consolidated in nature. Figure 3.90a and b depict a robust and

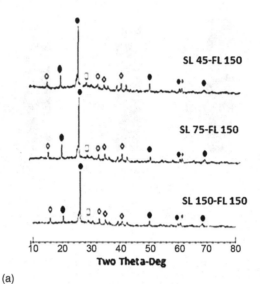

(a)

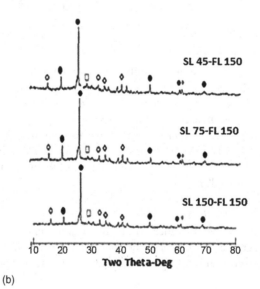

(b)

FIGURE 3.89 (a, b) XRD analysis of fly ash–slag-based AAC specimens for different fineness of slag.

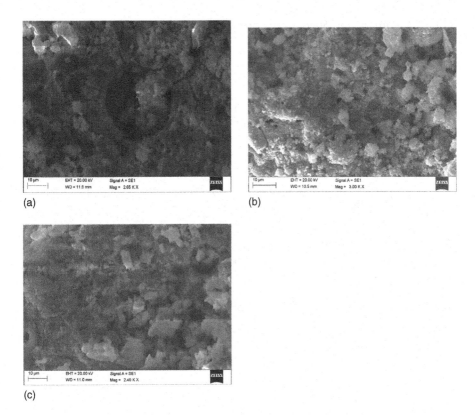

FIGURE 3.90 (a–c) SEM images for fly ash–slag AAC specimens for different fineness of slag.

cohesive microstructure. This is also due to the increase the intensity of the sodium alumino silicate gel as depicted in the XRD.

3.6.3.7 Water Sorptivity

Sorptivity is a material property which characterizes the tendency of a porous material to absorb and transmit water by capillary action. The measurement of sorptivity has primary importance in durability assessment. The durability of cementitious materials is largely influenced by the rate of ingress of detrimental ions such as chlorides, sulphates, etc. (water as a carrier). Hence, the knowledge of moisture transport through the cementitious material is of great importance.

3.6.3.7.1 Primary and Secondary Sorptivity

3.6.3.7.1.1 Effect of Slag Content

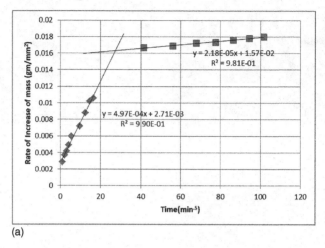

(a)

FIGURE 3.91a Sorptivity graph of (100-0) fly ash–slag AACs.

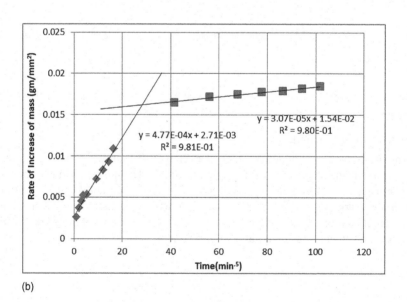

(b)

FIGURE 3.91b Sorptivity graph of (90-10) fly ash–slag AACs.

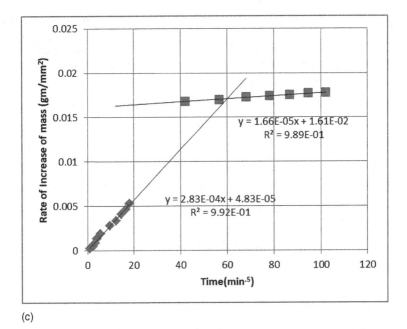

(c)

FIGURE 3.91c   Sorptivity graph of (85-15) fly ash–slag AACs.

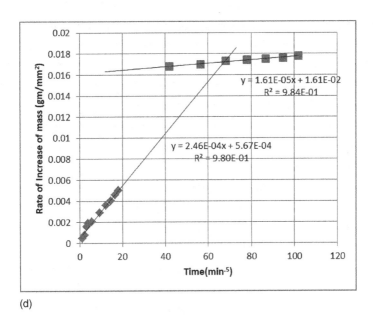

(d)

FIGURE 3.91d   Sorptivity graph of (70-30) fly ash–slag AACs.

The primary sorptivity and secondary sorptivity of four different mix specimens having 0%, 10%, 15%, and 30% slag content were recorded and are presented in Figures 3.92a, 3.92b, 3.92c, and 3.92d. The primary sorptivity was found to be 4.97 × 10^{-4}, 4.77 × 10^{-4}, 2.83 × 10^{-4}, and 2.46 × 10^{-4} for AAC paste specimens having a slag content of 0%, 10%, 15%, and 30%, respectively.

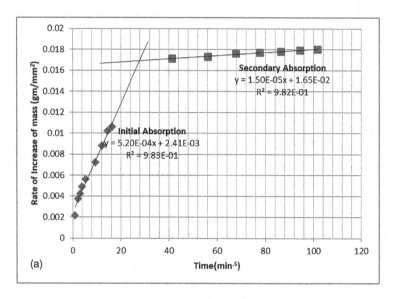

FIGURE 3.92a   Sorptivity graph of fly ash–slag AACs –4% Na$_2$O- A1 mix.

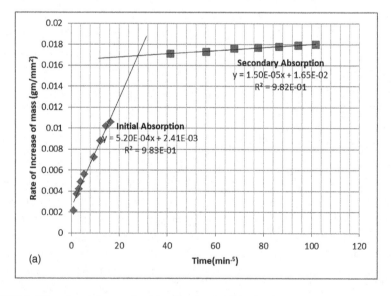

FIGURE 3.92b   Sorptivity graph of fly ash–slag AACs –6% Na$_2$O- A2 mix.

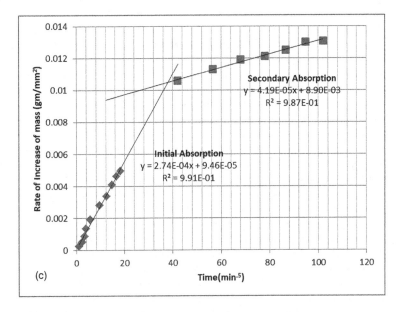

FIGURE 3.92c Sorptivity graph of fly ash–slag AACs −8% Na$_2$O- A3 mix.

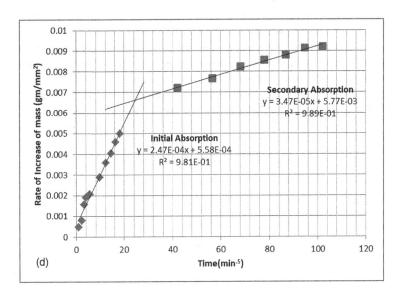

FIGURE 3.92d Sorptivity graph of fly ash–slag AACs −10% Na$_2$O- A4 mix.

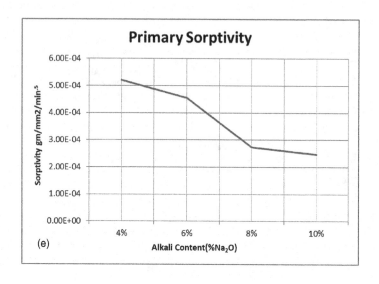

FIGURE 3.92e   Primary sorptivity vs. alkali content. (%Na$_2$O)

3.6.3.7.1.2 *Effect of Alkali Content*

The primary sorptivity values continued to decrease with the increase of alkaline content. The sorptivity values are shown in Table 3.21. The results can be correlated with the microstructure of the test specimen. It can be seen that the increase of alkaline content leads to a denser microstructure with fewer voids. As sorptivity is a measure of the capillary suction, it is dependent on the network of pores available in the gel matrix. The secondary rate of absorption or secondary sorptivity values were found to be 0.15 ×10^{-4}, 0.159 × 10^{-4}, 0.419 × 10^{-4} and 0.347 × 10^{-4} g/mm^2/min$^{0.5}$ for paste specimens having Na$_2$O content 4%, 6% and 8% and 10%, respectively [3.92 a,b,c,d,and e]. The secondary sorptivity values also followed the same trend as the primary sorptivity values, for example, it decreased with the increase of alkali content.

TABLE 3.21
Primary and Secondary Sorptivity of AAC Pastes

Mix ID	Primary Sorptivity (S$_p$) g/mm^2/$\sqrt{t}$	Secondary Sorptivity(S$_s$) g/mm^2/$\sqrt{t}$
A1-4%	S$_i$ = 5.2E-04	S$_s$ = 0.15E-04
A2-6%	S$_i$ = 4.54E-04	S$_s$ = 0.159E-04
A3-8%	S$_i$ = 2.74E-04	S$_s$ = 0.419E-04
A4-10%	S$_i$ = 2.47E-04	S$_s$ = 0.347E-04

3.6.3.7.1.3 Effect of Silica Content

The same trend is observed for secondary sorptivity. Adding SiO_2 beyond the 12% leads to a decrease of sorptivity values. Addition of silicate leads to the increased intensity of the dissolution process of polymerization but excess silicate hinders the later stages of the process. In Figure 3.93g, a correlation has been made between primary sorptivity, secondary sorptivity, and compressive strength. The relationship is non-linear in nature but a general impression, for example, a specimen with relatively high compressive strength will have comparatively lower primary and secondary sorptivity values.

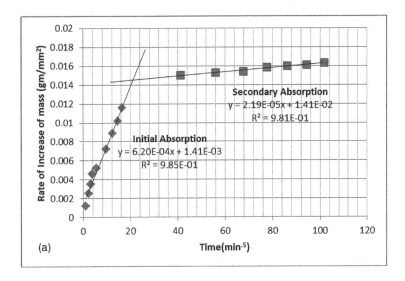

FIGURE 3.93a   Sorptivity of fly ash–slag AACs – B1-4% SiO_2.

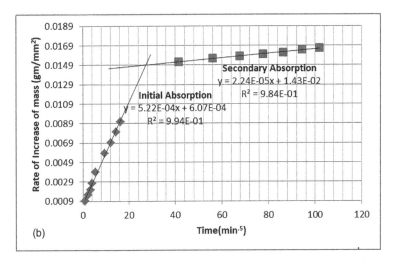

FIGURE 3.93b   Sorptivity of fly ash–slag AACs – B2-6% SiO_2.

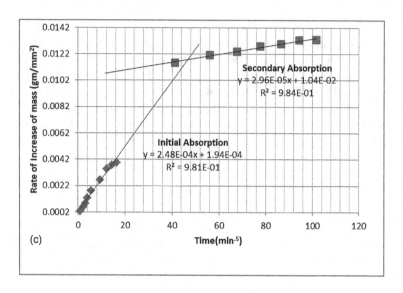

FIGURE 3.93c   Sorptivity of fly ash–slag AACs – B3-8% SiO_2.

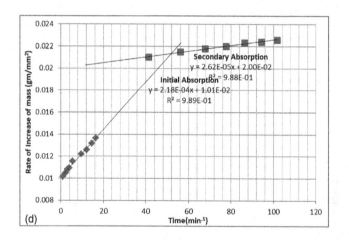

FIGURE 3.93d   Sorptivity of fly ash–slag AACs – B4-10% SiO_2.

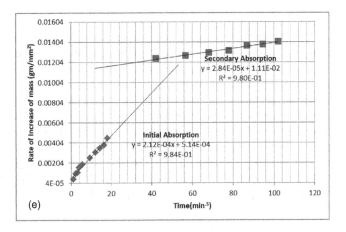

FIGURE 3.93e　Sorptivity of fly ash–slag AACs – B5-12% SiO_2.

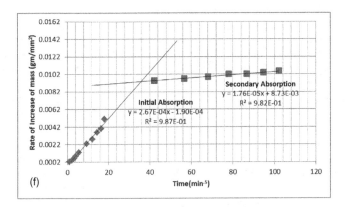

FIGURE 3.93f　Sorptivity of fly ash–slag AACs – B6-14% SiO_2.

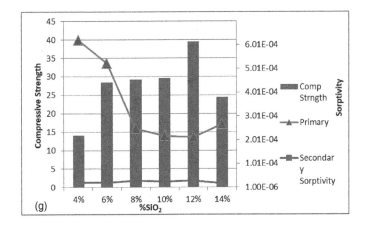

FIGURE 3.93g　Compressive strength vs. sorptivity for silica series.

3.6.3.7.1.4 Effect of Water Content

The values of the primary and secondary sorptivity as shown in Figures 3.94a, 3.94b, 3.94c, 3.94d, and 3.94e were observed to vary with the w/b ratio. The primary and secondary sorptivity values of the mix –C1-w/b = 0.35 are 5.76×10^{-4} g/mm²/min$^{.05}$ and 2.19×10^{-5}, respectively. These values decrease when the w/b ratio is increased to 0.38. The sorptivity values continue to increase with the increase in the w/b ratio beyond 0.38. The water does not directly participate in a polymeric reaction but it works as a medium of transportation of ions. Paucity of water can lead to an unavailability of ions, which hampers the polymerization reaction especially the dissolution stage. Also a less amount of water leads to a stiffer paste matrix which leads to incomplete compaction which in turn leads to more pores in the gel matrix. However, excess water gets trapped in the gel matrix and it evaporates when subjected to thermal curing leaving behind voids.

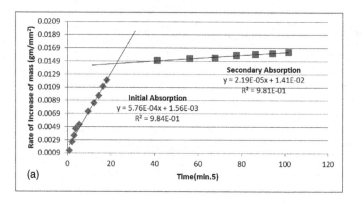

FIGURE 3.94a   Sorptivity of fly ash–slag AACs – C1- w/b = 0.35.

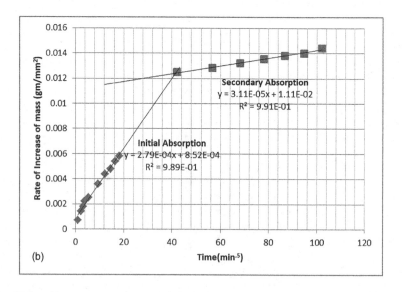

FIGURE 3.94b   Sorptivity of fly ash–slag AACs – C2- w/b = 0.38.

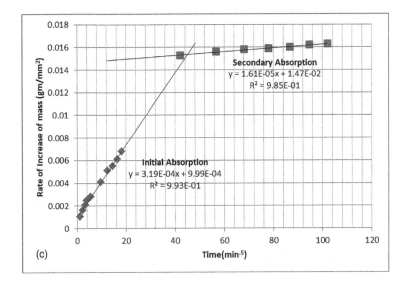

FIGURE 3.94c   Sorptivity of fly ash–slag AACs – C3-w/b = 0.41.

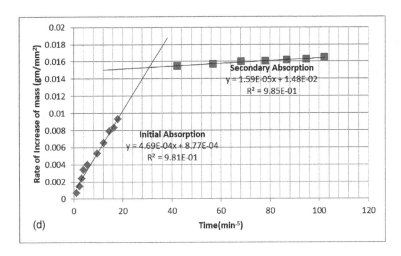

FIGURE 3.94d   Sorptivity of fly ash–slag AACs – C4-w/b = 0.44.

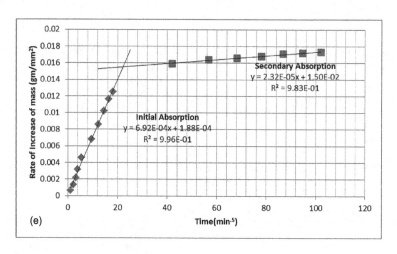

FIGURE 3.94e   Sorptivity of fly ash–slag AACs – C5-w/b = 0.47.

4 Durability of Alkali-Activated Paste and Mortar

4.1 PREAMBLE

This chapter presents the details of systematic experimental investigations conducted for studying the engineering properties of blended (fly Ash + blast furnace slag) AACs in the hardened state, exposed to elevated temperature, acids, and sulphate solutions, which have been presented and scientifically interpreted, to have a deeper understanding of the influence of different synthesis parameters on durability and microstructure.

Mineralogical/microstructure/elemental analysis and pore structure studies have been also carried out, using SEM/XRD/energy dispersive X-ray (EDAX)/FTIR/TGA and MIP techniques, for better understanding. Experimentation was conducted in the Concrete laboratory of Construction Engineering as well as in the laboratories of Metallurgical and Material Engineering Department of Jadavpur University, Kolkata, India, Indian Association for Cultivation of Science, UGC DAE-CSIR, Kolkata, and Central Glass and Ceramic Research Institute (CGCRI), Kolkata, India. For characterization of fly ash-based AACs, mineralogical/microstructure/elemental analysis and pore structure studies were carried out using SEM/XRD/EDAX/FTIR/TGA and MIP techniques. The laboratory tests were conducted as per relevant Indian standard codes and ASTM standards. Studies were performed on the effect of elevated temperature on physico-mechanical properties, microstructure, and mineralogical changes on blended AACs up to 1000°C. Studies were also performed to assess the physico-mechanical properties, microstructure, and mineralogical changes of blended AACs when exposed to sulphuric acid and nitric acid of varying concentrations as well as when exposed to the magnesium sulphate and sodium sulphate solution of varying concentrations also.

4.2 EXPERIMENTAL INVESTIGATION OF FLY ASH–SLAG BLENDED AACS

4.2.1 STUDY ON PHYSICO-MECHANICAL PROPERTIES OF AACs EXPOSED TO ELEVATED TEMPERATURE

The experimental program was undertaken to study the effect of slag content, alkali content (Na_2O), silica content ($\%SiO_2$), and water/binder Ratio on thermo-mechanical and microstructural properties of AACs exposed to elevated temperature. Four series of experiments were conducted (slag series, alkali series, silica series, and water/binder Series). In slag series, the percentage of slag was varied (0%, 10%, 15%, and 30%).

FIGURE 4.1 Test specimens exposed to elevated temperature in a muffle furnace.

In alkali series, the effect of alkali content on engineering properties of AACs was studied. AAC specimens were prepared by varying the % (Na_2O) from 4% to 12% and keeping constant %SiO_2 equal to 8% slag content fixed at 30%. In the second series (silica series), the effect of SiO_2 was studied on specimens prepared by varying the SiO_2 content from 4% to 14% by the weight of the source material by keeping % (Na_2O) and slag % constant. The specimens were heated in a muffle furnace (Figure 4.1) for four hours at 400°C, 600°C, and 800°C and air-cooled. The changes in compressive strength, bulk density, apparent porosity, water absorption loss of weight, and microstructure of AAC paste specimens have been reported. TGA was conducted to study the loss of mass of specimens during transient heating. The mineralogical and microstructural changes were studied using XRD and SEM/energy dispersive X-ray (EDX) analysis.

4.2.2 Study on AACs Exposed to Acid Attack

The experimental program was undertaken to study the resistance of AACs on exposure to different concentrations of acid solutions. To study the effect of concentration of acid solution, 6% and 8% concentrated sulphuric and nitric acid have been used for exposing the AAC paste specimens for exposure up to 16 weeks. The test specimens were immersed in an acid solution after two days of oven curing at 85°C. In order to maintain the concentration, the solution was replaced every month and stirred every week. During the exposure, regular investigations on physical changes, weight changes, and residual strength were performed at predetermined intervals. An optical microscope was used to observe surface texture and corrosion of the surface. The pH of acid solution was monitored regularly using a digital pH meter. For measuring change in weight due to acid exposure, the specimens were taken out of the solution, and they were wiped dry. Residual compressive strength was based on the initial 28 days' compressive strength of specimens. At specific intervals, the samples

FIGURE 4.2   AAC specimens exposed to sulphuric acid, nitric acid, sodium sulphate, and magnesium sulphate.

were taken out from the exposure solution, wiped, and then tested in a compression-testing machine. At selected intervals, samples were tested for mineralogical and microstructural changes using SEM/EDX analysis.

4.2.3 STUDY ON AACs EXPOSED TO SULPHATE ATTACK

In this experimental program, sulphate resistance of AACs was studied for different concentrations of magnesium sulphate solutions. For paste specimens, 6% and 8% sodium sulphate and magnesium sulphate solution were used to study the effect of concentrations, for an exposure duration of 16 weeks. The test specimens were immersed in an acid solution as shown in Figure 4.2 after 28 days of water curing. In order to maintain the concentration, the solution was replaced every month and stirred every week. During the exposure, regular investigations on physical changes, change in pH, weight changes, and residual strength were monitored at predetermined intervals. An optical microscope was used to observe the surface texture for physical change. The pH of the sulphate solution was monitored regularly using a digital pH meter. For measuring change in weight due to sulphate exposure, the specimens were taken out of the solution and wiped dry. Residual compressive strength was based on the initial 28 days' compressive strength of specimens. At specific intervals, the samples were taken out from the exposure solution, wiped, and then tested in a compression-testing machine. At selected intervals, samples were tested for mineralogical and microstructural changes using SEM/EDX analysis.

4.2.4 STUDY ON PROPERTIES OF FLY ASH–SLAG BLENDED AACs EXPOSED TO ELEVATED TEMPERATURE

4.2.4.1 Mix Proportions

In this study, four series of AAC test specimens were prepared.

In the first test series (Refer Table 4.1), the slag content was varied from 0% to 60%. The alkali and SiO_2 content in the mix has been kept constant at 8% and 8%, respectively. In the second test series (Refer Table 4.2), the alkali content was varied from 4% to 10% and the silica content and water content was kept constant at 8% and 0.38, respectively. In the third test series (Refer Table 4.3), the silica content was varied from 4% to 14% and the Na_2O content and water content was kept constant at 8% and 0.38, respectively. In the fourth test series (Refer Table 4.4), the water to

TABLE 4.1

Details of AAC Mix for Slag Series (Exposed to Elevated Temperature)

AAC Mix Details

Mix ID	% Na_2O in Activator (a)*	% SiO_2 in Activator (b)*	Water/ Binder Ratio (c)*	Fly Ash (gm)	Slag (gm)	Sodium Silicate (gm)	Extra Water Added (gm)	Sodium Hydroxide (gm)
100–0	8	8	0.38	1000	0	301.89	165.55	74.29
90–10	8	8	0.38	900	100	301.89	165.55	74.29
85–15	8	8	0.38	850	150	301.89	165.55	74.29
70–30	8	8	0.38	700	300	301.89	165.55	74.29
50–50	8	8	0.38	500	500	301.89	165.55	74.29
40–60	8	8	0.38	400	600	301.89	165.55	74.29

Notes: (a)*, (b)*, and (c)* by weight of source material.
Specimen calculation available in Appendix.

TABLE 4.2

Details of AAC Mix for Alkali Series (Exposed to Elevated Temperature)

AAC Mix Details

Mix ID	% Na_2O in Activator (a)*	% SiO_2 in Activator (b)*	Water/ Binder Ratio (c)*	Fly Ash (gm)	Slag (gm)	Sodium Silicate (gm)	Extra Water Added (gm)	Sodium Hydroxide (gm)
A1–4%	4	8	0.38	700	300	301.89	177.52	21.08
A2–6%	6	8	0.38	700	300	301.89	171.53	47.69
A3–8%	8	8	0.38	700	300	301.89	165.55	74.29
A4–10%	10	8	0.38	700	300	301.89	159.56	100.90

Note: (a)*, (b)*, and (c)* by weight of source material (fly ash + slag).

binder ratio was varied from 4% to 14% and the Na_2O content and silica content was kept constant at 8% and 0.38, respectively. Depending on the quantity of water available in the NaOH pellets and Na_2SiO_3 solution, extra water has been added into the activator solution to make up the required water to fly ash–slag ratio.

At the age of 28 days, the AAC paste specimens of each test series were subjected to elevated temperatures of 400°C, 600°C, and 800°C in a muffle furnace as shown in Figure 4.1. The temperature in the muffle furnace was increased at a rate of 5°C per minute starting from room temperature [Figure 4.3]. After attaining the target temperature in a furnace, it was maintained for an additional four hour before the furnace was shut down to allow the specimens to cool down to room temperature in the furnace. The unexposed counterparts were left undisturbed at room temperature until testing for comparative studies.

TABLE 4.3
Details of AAC Mix for Silica Series (Exposed to Elevated Temperature)

AAC Mix Details

Mix ID	% Na$_2$O in Activator (a)*	% SiO$_2$ in Activator (b)*	Water/ Binder Ratio (c)*	Fly Ash (gm)	Slag (gm)	Sodium Silicate (gm)	Extra Water Added (gm)	Sodium Hydroxide (gm)
B1–4%	8	4	0.38	700.00	300	150.94	260.80	90.36
B2–6%	8	6	0.38	700.00	300	226.42	213.18	82.32
B3–8%	8	8	0.38	700.00	300	301.89	165.55	74.29
B4–10%	8	10	0.38	700.00	300	377.36	117.92	66.26
B5–12%	8	12	0.38	700.00	300	452.83	70.29	58.23
B6–14%	8	14	0.38	700.00	300	528.30	22.67	50.20

Note: (a)*, (b)*, and (c)* by weight of source material (fly ash + slag).

TABLE 4.4
Details of AAC Mix for Water/Binder Series (Exposed to Elevated Temperature)

AAC Mix Details

Mix ID	% Na$_2$O in Activator (a)*	% SiO$_2$ in Activator (b)*	Water/Binder Ratio (c)*	Fly Ash (gm)	Slag (gm)	Sodium Silicate (gm)	Extra Water Added (gm)	Sodium Hydroxide (gm)
C1-0.35	8	8	0.35	700	300	301.89	135.55	74.29
C2-0.38	8	8	0.38	700	300	301.89	165.55	74.29
C3-0.41	8	8	0.41	700	300	452.83	100.29	58.23
C4-0.44	8	8	0.44	700	300	452.83	130.29	58.23
C5-0.47	8	8	0.47	700	300	452.83	160.29	58.23

Note: (a)*, (b)*, and (c)* by weight of source material (fly ash + slag).

4.2.4.2 Tests Conducted
Compressive strength, bulk density, apparent porosity, weight loss measurements, TGA/differential thermal analysis (DTA), XRD analysis, and SEM/EDAX were carried out.

4.2.4.3 Study on the Properties of AAC Specimens Exposed to Acid Attack
In this study, four series of AAC specimens were prepared.

4.2.4.3.1 Test Series-1: Effect of Slag Content
In this series, AAC paste specimens were cast by varying slag content from 10% to 30% and keeping Na$_2$O content at 8% and SiO$_2$ content at 8%, constant. The water

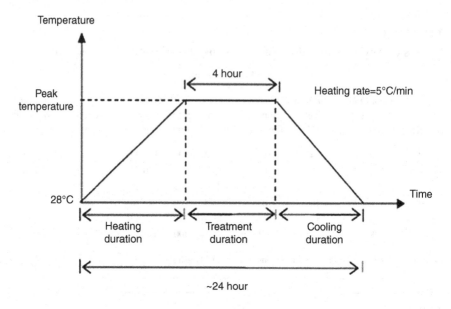

FIGURE 4.3 Heating regime of AAC specimens exposed to elevated temperature.

TABLE 4.5
Test Series-1: Effect of Slag Content

			AAC Mix Details					
Mix ID	% Na$_2$O in Activator (a)*	% SiO$_2$ in Activator (b)*	Water/Binder Ratio (c)*	Fly Ash (gm)	Slag (gm)	Sodium Silicate (gm)	Extra Water Added (gm)	Sodium Hydroxide (gm)
90–10	8	8	0.38	900	100	301.89	165.55	74.29
85–15	8	8	0.38	850	150	301.89	165.55	74.29
70–30	8	8	0.38	700	300	301.89	165.55	74.29

Note: (a)*, (b)*, and (c)* by weight of source material (fly ash + slag).

to binder ratio was kept constant equal to 0.38. The mix proportion of AAC mixes is provided in Table 4.5.

4.2.4.3.2 Test Series-2: Effect of Alkali Content

In this series, AAC paste specimens were cast by varying alkali content (Na$_2$O) from 4% to 8% and keeping constant SiO$_2$ content equal to 10%. The water to binder ratio was kept constant equal to 0.38. The changes in the AAC mix were obtained by varying the amount of sodium hydroxide sodium silicate solution and water (if required). The mix proportion of AAC mixes is presented in Table 4.6.

TABLE 4.6
Details of AAC Mix for Alkali Series

AAC Mix Details

Mix ID	% Na$_2$O in Activator (a)*	% SiO$_2$ in Activator (b)*	Water/Binder Ratio (c)*	Fly Ash (gm)	Slag (gm)	Sodium Silicate (gm)	Extra Water Added (gm)	Sodium Hydroxide (gm)
A1	4.0	10.00	0.38	700	300	377.36	129.89	13.05
A2	6.0	10.00	0.38	700	300	377.36	123.91	39.66
A3	8.0	10.00	0.38	700	300	377.36	117.92	66.26

Note: (a)*, (b)*, and (c)* by weight of source material (fly ash + slag).

TABLE 4.7
Details of AAC Mix for Silica Series

AAC Mix Details

Mix ID	% Na$_2$O in Activator (a)*	% SiO$_2$ in Activator (b)*	Water/Binder Ratio (c)*	Fly Ash (gm)	Slag (gm)	Sodium Silicate (gm)	Extra Water Added (gm)	Sodium Hydroxide (gm)
B1–4%	8	4	0.38	700	300	150.94	260.80	90.36
B2–6%	8	6	0.38	700	300	226.42	213.18	82.32
B3–8%	8	8	0.38	700	300	301.89	165.55	74.29
B4–10%	8	10	0.38	700	300	377.36	117.92	66.26
B5–12%	8	12	0.38	700	300	452.83	70.29	58.23
B6–14%	8	14	0.38	700	300	528.30	22.67	50.20

Note: (a)*, (b)*, and (c)* by weight of source material (fly ash + slag).

4.2.4.3.3 Test Series-3: Effect of Silica Content

In this series, AAC specimens were cast with varying silica content (SiO$_2$) from 4% to 14% by keeping constant alkali content (Na$_2$O) equal to 8%. The water to binder ratio was kept constant equal to 0.38. The changes in AAC mix were obtained by varying the amount of sodium hydroxide, sodium silicate solution, and water (if required). The mix proportions are presented in Table 4.7.

4.2.4.3.2 Test Series-4: Effect of Water Content

In this series, AAC specimens were cast by varying the water to binder ratio from 0.35 to 0.47 by keeping constant alkali content (Na$_2$O) equal to 8% and silica content (SiO$_2$) 10%. The changes in the AAC mix were obtained by varying the amount of sodium hydroxide, sodium silicate solution and water (if required). The mix proportions are presented in Table 4.8.

TABLE 4.8
Details of AAC Mix for Water/Binder Series

AAC Mix Details

Mix ID	% Na$_2$O in Activator (a)*	% SiO$_2$ in Activator (b)*	Water/Binder Ratio (c)*	Fly Ash (gm)	Slag (gm)	Sodium silicate (gm)	Extra water added (gm)	Sodium Hydroxide (gm)
C1-0.35	8	10	0.35	700	300	377.36	88.37	66.20
C2-0.38	8	10	0.38	700	300	377.36	118.37	66.20
C3-0.41	8	10	0.41	700	300	377.36	148.37	66.20
C4-0.44	8	10	0.44	700	300	377.36	178.37	66.20
C5-0.47	8	10	0.47	700	300	377.36	208.37	66.20

Note: (a)*, (b)*, and (c)* by weight of source material (fly ash + slag).

TABLE 4.9
Test Series-1: Effect of Slag Content

AAC Mix Details

Mix ID	% Na$_2$O in Activator (a)*	% SiO$_2$ in Activator (b)*	Water/Binder Ratio (c)*	Fly Ash (gm)	Slag (gm)	Sodium Silicate (gm)	Extra Water Added (gm)	Sodium Hydroxide (gm)
90-10	8	8	0.38	900	100	301.89	165.55	74.29
85-15	8	8	0.38	850	150	301.89	165.55	74.29
70-30	8	8	0.38	700	300	301.89	165.55	74.29

Note: (a)*, (b)*, and (c)* by weight of source material (fly ash + slag).

4.2.4.4 Test Conducted

Compressive strength, physical appearance, weight loss measurements, XRD analysis, and SEM/EDAX analysis were carried out.

4.2.5 STUDY ON PROPERTIES OF BLENDED AACs EXPOSED TO SULPHATE ATTACK

4.2.5.1 Mix Proportions

In this study, four series of AAC specimens were prepared. For exposure to sulphate solution, paste specimens have been used. The details of the specimens are given in Tables 4.2.5.1 a–4.2.5.1 d.

4.2.5.1.1 Test Series-1: Effect of Slag Content

In this series, AAC paste specimens were cast by varying slag content from 10% to 30%, keeping Na$_2$O at 8% and SiO$_2$ content equal to 10%. The water to binder ratio was kept constant equal to 0.38. The mix proportion of AAC mix is presented in Table 4.9.

TABLE 4.10
Details of AAC Mix for Alkali Series

AAC Mix Details

Mix ID	% Na$_2$O in Activator (a)*	% SiO$_2$ in Activator (b)*	Water/Binder Ratio (c)*	Fly Ash (gm)	Slag (gm)	Sodium Silicate (gm)	Extra Water Added (gm)	Sodium Hydroxide (gm)
A1	4	10	0.38	700	300	377.36	129.89	13.05
A2	6	10	0.38	700	300	377.36	123.91	39.66
A3	8	10	0.38	700	300	377.36	117.92	66.26

Note: (a)*, (b)*, and (c)* by weight of source material (fly ash + slag).

TABLE 4.11
Details of AAC Mix for Silica Series

AAC Mix Details

Mix ID	% Na$_2$O in Activator (a)*	% SiO$_2$ in Activator (b)*	Water/Binder Ratio (c)*	Fly Ash (gm)	Slag (gm)	Sodium Silicate (gm)	Extra Water Added (gm)	Sodium Hydroxide (gm)
B1–4%	8	4	0.38	700	300	150.94	260.80	90.36
B2–6%	8	6	0.38	700	300	226.42	213.18	82.32
B3–8%	8	8	0.38	700	300	301.89	165.55	74.29
B4–10%	8	10	0.38	700	300	377.36	117.92	66.26
B5–12%	8	12	0.38	700	300	452.83	70.29	58.23
B6–14%	8	14	0.38	700	300	528.30	22.67	50.20

Note: (a)*, (b)*, and (c)* by weight of source material (fly ash + slag).

4.2.5.1.2 Test Series-2: Effect of Alkali Content

In this series, AAC paste specimens were cast by varying alkali content (Na$_2$O) from 4% to 8% and keeping constant SiO$_2$ content equal to 10%. The water to binder ratio was kept constant equal to 0.38. The changes in the AAC mix were obtained by varying the amount of sodium hydroxide, sodium silicate solution, and water (if required). The mix proportion of AAC mix is presented in Table 4.10.

4.2.5.1.3 Test Series-3: Effect of Silica Content

In this series, AAC specimens were cast by varying silica content from 4% to 14% and keeping constant alkali content (Na$_2$O) equal to 8%. The water to binder ratio was kept constant equal to 0.38. The changes in the AAC mix were obtained by varying the amount of sodium hydroxide, sodium silicate solution, and water (if required). The mix proportion of the AAC mix is presented in Table 4.11.

TABLE 4.12

Details of AAC Mix for Water/Binder Series

	AAC Mix Details							
Mix ID	% Na_2O in Activator (a)*	% SiO_2 in Activator (b)*	Water/Binder Ratio (c)*	Fly Ash (gm)	Slag (gm)	Sodium Silicate (gm)	Extra Water Added (gm)	Sodium Hydroxide (gm)
C1-0.35	8	10	0.35	700	300	377.36	88.37	66.20
C2-0.38	8	10	0.38	700	300	377.36	118.37	66.20
C3-0.41	8	10	0.41	700	300	377.36	148.37	66.20
C4-0.44	8	10	0.44	700	300	377.36	178.37	66.20
C5-0.47	8	10	0.47	700	300	377.36	208.37	66.20

Note: (a)*, (b)*, and (c)* by weight of source material (fly ash + slag).

4.2.5.1.4 Test Series-4: Effect of Water Content

In this series, AAC specimens were cast by varying water binder content from 0.35 to 0.47 and keeping constant alkali content (Na_2O) equal to 8% and silica content 10%. The changes in the AAC mix were obtained by varying the amount of sodium hydroxide, sodium silicate solution, and water (if required) in the activating solution. The mix proportion of AAC mix is presented in Table 4.12.

4.2.5.2 Test Conducted

Compressive strength, physical appearance, weight loss measurements, XRD analysis, and SEM/EDAX analysis were performed.

4.3 RESULTS AND DISCUSSION

4.3.1 BLENDED AACs EXPOSED TO ELEVATED TEMPERATURE

In this study, four different series of AAC specimens were prepared. Slag, sodium hydroxide, sodium silicate, and water contents were varied. The specimens were exposed to the temperature level of 400°C, 600°C, and 800°C. Visual inspection and residual compressive strength, apparent porosity, bulk density, and water absorption measurements were performed. Microstructural study by SEM was performed. The TGA/DTA test was also conducted.

4.3.1.1 Test Series-1: Effect of Slag Content

In this series, AAC specimens were prepared by varying slag content. The slag was varied from 0% to 30% by keeping constant Na_2O and SiO_2 content of 8%. The water to binder ratio was kept constant equal to 0.38. Oven curing was carried out for 48 hours, before exposing the specimens to elevated temperature (400°C, 600°C, and 800°C) for four-hour duration in a muffle furnace. The mix proportion of AAC mix is presented in Table 4.1.

4.3.1.2 Test Series-2: Effect of Alkali Content

In this series, AAC specimens were prepared by varying alkali content (Na_2O). The % (Na_2O) by weight of (fly ash + slag) was varied from 4% to 10% by keeping constant SiO_2 content of 8%. The water to binder ratio was kept constant equal to 0.38. Oven curing was carried out for 48 hours, before exposing the specimens to elevated temperature (400°, 600°, and 800°C) for four-hour duration in a muffle furnace. The mix proportion of AAC mix is presented in Table 4.2.

4.3.1.3 Test Series-3: Effect of Silica Content

In this series, AAC specimens were prepared with varying silica content (% SiO_2) from 4% to 14% by keeping constant alkali content (%Na_2O) = 8%. The water to binder ratio was kept constant equal to 0.38. Oven curing was carried out for 48 hours, before exposing the specimens to elevated temperature (400°, 600°, and 800°C) for four-hour duration in a muffle furnace. The mix proportion of AAC mix is presented in Table 4.3.

4.3.1.4 Test Series-4: Effect of Water Content

In this series, AAC specimens were prepared by varying water content. The water/binder ratio was varied from 0.35 to 0.47 by keeping constant Na_2O and SiO_2 content at 8%. Oven curing was carried out for 48 hours, before exposing the specimens to elevated temperature (400°C, 600°C, and 800°C) for four-hour duration in a muffle furnace. The mix proportion of AAC mix is presented in Table 4.4.

4.3.1.5 Visual Observations

Figure 4.4 indicates the change in surface texture and color after exposing the test specimens to elevated temperature. The specimen color remains more or less unchanged compared to the original unexposed specimens up to a temperature of 400°C. From 400°C to 600°C, a few red patches were noticed on the samples. At 800°C, most of the surfaces turned red.

4.3.1.6 Bulk Density, Apparent Porosity, and Water Absorption

4.3.1.6.1 Effect of Slag Content

Figures 4.5–4.7 indicate the change in the bulk density, apparent porosity, and water absorption values of 100-0 specimens, respectively; it was seen that the values improve with elevated temperature exposure up to 800°C but this trend does not hold for specimens containing slag which continues to deteriorate upto 800°C. As the calcium silicate gel decomposes at around 690°C, the 90-10, 85-15, 70-30, 50-50, and 40-60 specimens deteriorate with elevated temperature exposure from 600°C to 800°C. The fly ash-based aluminosilicate polymer gel undergoes further polymerization and forms new gel products at temperature beyond 600°C to 800°C, thus contributing to the strength. Thus, the rate of decrease of strength beyond 600°C to 800°C is less for specimens 90-10, 85-15, and 70-30. Specimens 100-0, 90-10, and 85-15 having the aluminosilicate gel as the major binding gel performed better relative to 70-30 which loses as the amount of calcium silicate gel is much higher compared to other mixes.

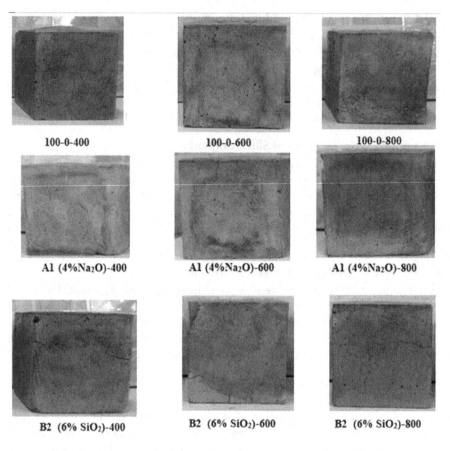

FIGURE 4.4   Photograph image of AAC paste specimens exposed to elevated temperature (follow specimen marking as indicated before in the tables).

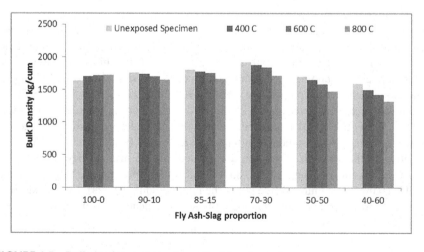

FIGURE 4.5   Bulk density vs. temperature level for slag series.

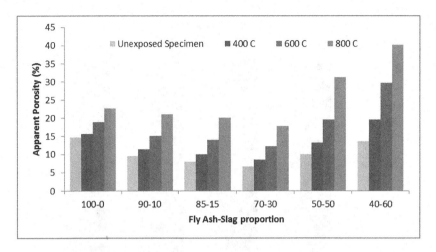

FIGURE 4.6 Apparent porosity vs. temperature level for slag series.

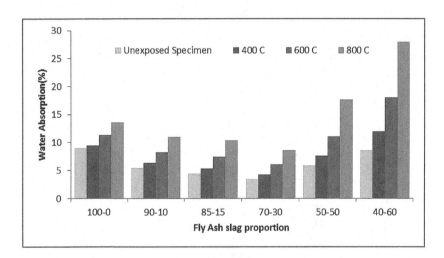

FIGURE 4.7 Water absorption vs. temperature level for slag series.

4.3.1.6.2 Effect of Alkali Content

The effect of alkali content on bulk density of AAC paste specimens exposed to 400°C, 600°C, and 800°C is presented in Figure 4.8. The bulk density of AAC specimens depends on the alkali content of the mix and exposure temperature. The maximum bulk density of 1750.15, 1662.64, and 1629.39 kg/m^3 was obtained for 8% alkali content at 400°C, 600°C, and 800°C, respectively. For 400°C, the minimum bulk density recorded was 1572.55 kg/m^3 for the A4 specimen. For 600°C, the minimum bulk density of 1410.78 kg/m^3 was obtained for A1 (4%Na$_2$O). For 800°C, the minimum bulk density of 1410.78 kg/m^3 was obtained for A1. The lowering of bulk density up to 600°C takes place due to the evaporation of free water present in the specimen. There is a breakdown of calcium aluminum silicate hydrate, which causes a decrease in the bulk density beyond 600°C to 800°C. Figures 4.9 and 4.10

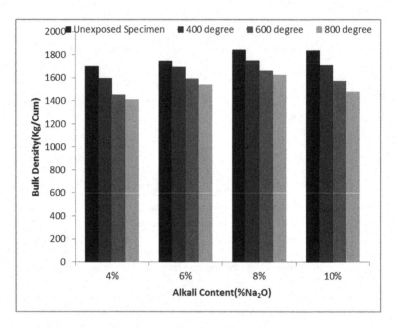

FIGURE 4.8   Bulk density vs. temperature level for alkali series. (Na_2O = 4, 6, 8, and 10%)

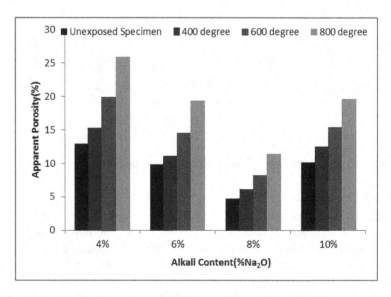

FIGURE 4.9   Apparent porosity vs. temperature level for alkali series. (Na_2O = 4, 6, 8, and 10%)

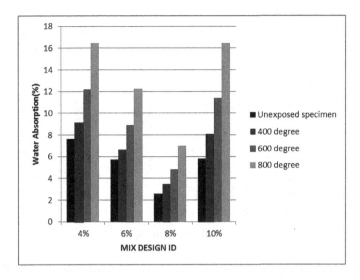

FIGURE 4.10 Water absorption vs. temperature level for alkali series. (Na_2O 4%, 6%, 8%, and 10%)

show the effect of alkali content on apparent porosity and water absorbtion of the AAC specimen at various temperatures. It is seen that the apparent porosity increases with the temperature exposure level. The apparent porosity values for A1(4%Na_2O), A2(6%Na_2O), A3(8%Na_2O), and A4(4%Na_2O) for 400°C are 15.28%, 11.08%, 6.12%, and 12.48%, respectively, and further increase to 19.86%, 14.51%, 8.2%, and 15.48% for 600°C, respectively. The apparent porosity values continue to increase when the temperature goes beyond 600°C. The apparent porosity values of A1, A2, A3, and A4 at 800°C are 25.82%, 19.3%, 11.48%, and 19.5%, respectively.

4.3.1.6.3 Effect of Silica Content

The effect of silica content on bulk density of AAC paste specimens at 400°C, 600°C, and 800°C is presented in Figure 4.11. The bulk density of AAC specimens depends on the silica content of the mix and exposure temperature. The maximum bulk density at 400°C, 600°C, and 800°C was 1692.19, 1677.08, and 1633.33 kg/m^3 for 12% silica content, respectively. The minimum bulk density was 1516.71, 1395.38, and 1352.11 kg/m^3 at 4% SiO_2 content. Figures 4.12 and 4.13 show the effect of silica content on apparent porosity and water absorption of the AAC specimen at various temperatures. It is observed that the apparent porosity and water absorption values increase almost linearly with increasing temperature level up to 600°C. Thereafter, there was also a substantial increase in apparent porosity between 600°C and 800°C, as shrinkage was more predominant in these ranges of

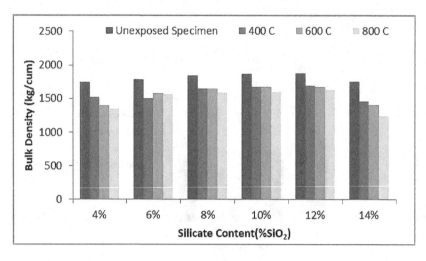

FIGURE 4.11 Bulk density vs. temperature level for silica series. (4%, 6%, 8%, 10%, 12%, and 14% SiO$_2$)

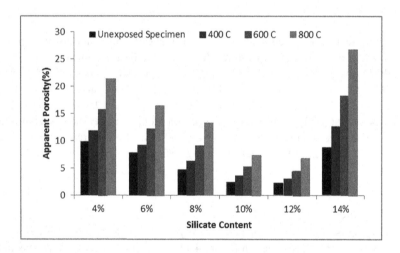

FIGURE 4.12 Apparent porosity vs. temperature level for silica series. (4%, 6%, 8%, 10%, 12%, and 14% SiO$_2$)

temperature. The maximum and minimum apparent porosity for specimens exposed to 800°C was found to be 26.51% and 6.75% for specimens having silica content of 4% and 12%, respectively. It can be also seen that the apparent porosity decreases with the increase in silica content from 0% to 12% but decreases beyond 12%. It was also observed that apparent porosity increases with increasing exposure temperature level.

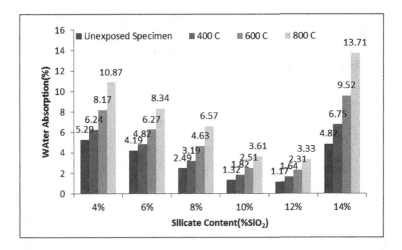

FIGURE 4.13   Water absorption vs. temperature effect of silica content. (4%, 6%, 8%, 10%, 12%, and 14% SiO_2)

4.3.1.6.4 Effect of Water Content

The effect of the water/binder ratio on bulk density of AAC paste specimens at 400°C, 600°C, and 800°C is presented in Figure 4.14. The bulk density of the AAC specimen is clearly dependent on the water/binder ratio of the mix and exposure temperature level. The maximum bulk density at 400°C, 600°C, and 800°C was 1726.56, 1736, and 1680 kg/m³ obtained for a water/binder ratio of 0.38, respectively. The minimum bulk density was 1371.36, 1350.69, and 1307.12 kg/m³ at a

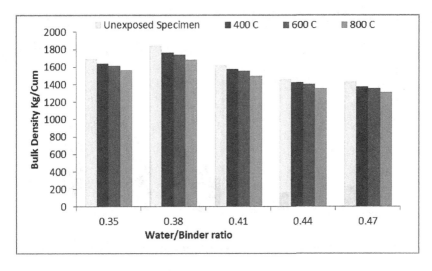

FIGURE 4.14   Bulk density vs. temperature level for water–binder series.

water/binder ratio of 0.47. The apparent porosity values at 400°C, 600°C, and 800°C were 3.08%, 4%, and 5.48%, respectively. (Figure 4.15) The maximum water absorption values at 400°C, 600°C, and 800°C were 5.51%, 7%, and 9.38%, respectively, for a water/binder ratio of 0.38 (Figure 4.16).

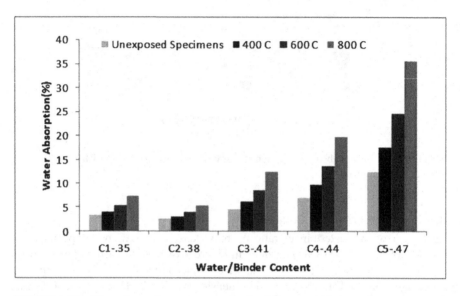

FIGURE 4.15 Water absorption vs. temperature level for water–binder series.

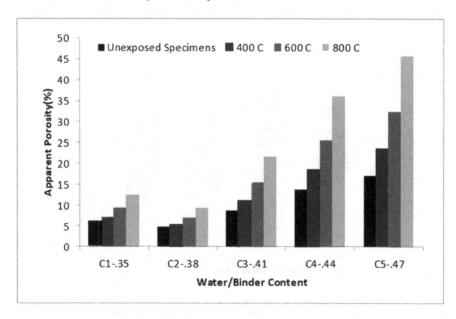

FIGURE 4.16 Apparent porosity vs. temperature level for water–binder series.

4.3.1.7 Weight Loss Measurements by TGA/DTA Analysis

The TGA technique was used to measure the changes in weight loss of AAC paste specimens of alkali and water/binder series as a function of time. The TGA measurements were performed between 28°C and 1000°C at a constant heating rate of 100°C per minute in inert media with nitrogen purge rate of 150 ml/min using a Pyris Diamond Perkin-Elmer TGA instrument and a platinum crucible with alpha alumina powder as a reference.

4.3.1.7.1 Effect of Alkali Content

Figures 4.17 presents the weight loss percentage for AAC paste specimens having varying alkali content when exposed to a temperature of approximately 1000°C. The TGA curve shows the total weight loss of all the AAC specimens with varying alkali content (Figures 4.18–4.21). The weight loss up to a temperature of 1000°C was recorded. The lowest weight loss was recorded by the specimen with 8% Na_2O with a weight retention of 86.44%. From the TGA graphs, it is seen that 12% of weight loss occurred for specimens with Na_2O up to a temperature of 400°C but, on the other hand, it took 700°C for specimen A3 to a loss of 12% of its weight, and it is also noticed that it recorded the lowest weight loss among the specimens. In the A2 specimen, 16% mass loss occurred at 400°C and A5 recorded a much higher mass loss of 18% at 400°C. This mass loss is closely related to the rigidity of the hybrid alumina-calcium-silicate hydrate gel. The higher the alkaline content, the higher the rigidity of the gel structure and the higher the energy requirement for its disintegration which leads to the lesser mass loss at elevated temperature compared to a specimen with a lower alkali content.

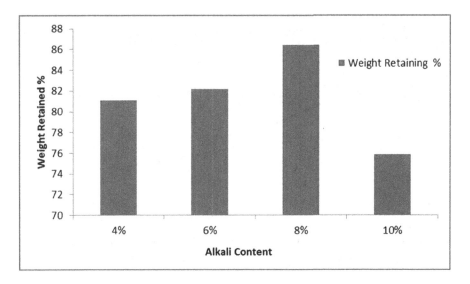

FIGURE 4.17   Alkali content (% Na_2O) vs. weight retained.

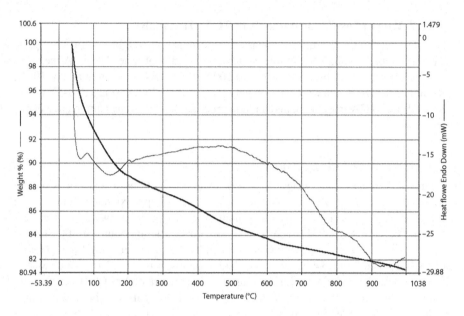

FIGURE 4.18   TGA/DTA curve for the A1 (4% Na$_2$O) specimen. (alkali series)

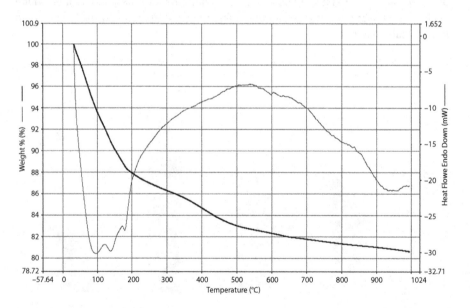

FIGURE 4.19   TGA/DTA curve for the A2 (6% Na$_2$O) specimen. (alkali series)

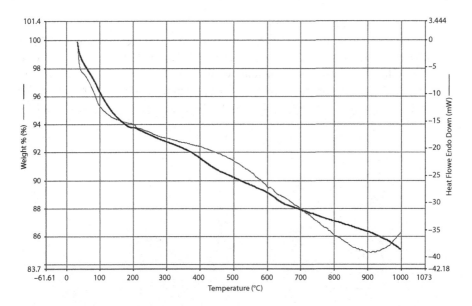

FIGURE 4.20 TGA/DTA curve for the A3 (8% Na$_2$O) specimen. (alkali series)

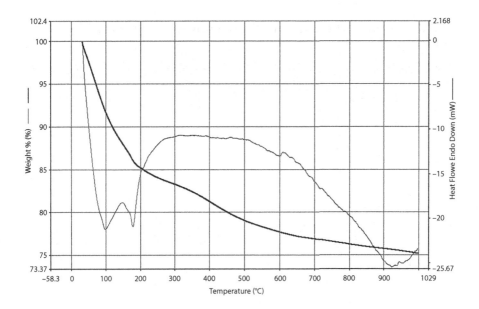

FIGURE 4.21 TGA/DTA curve for the A4 (10% Na$_2$O) specimen. (alkali series)

4.3.1.7.2 Effect of Water Content

Figures 4.22–4.26 present typical TGA curves for AAC paste specimens with different water/binder ratios having an alkali content of 8% Na_2O and 8% SiO_2 from 32.25°C to 999.75°C. The TGA curve shows the total weight loss of all the AAC specimens with water/binder content. The weight loss up to a temperature of 1000°C was recorded. The lowest weight loss was recorded by the specimen with 0.35 water/binder ratio with a

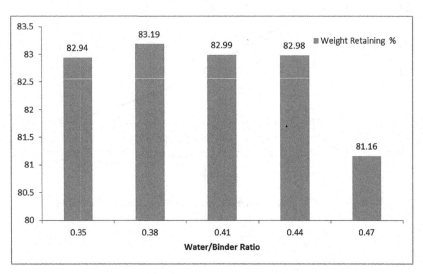

FIGURE 4.22 TGA/DTA curve for the C1 (water–binder ratio = 0.35) specimen. (water–binder series).

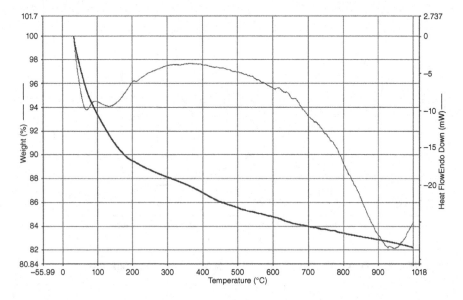

FIGURE 4.23 TGA/DTA curve for the C2 (water–binder ratio = 0.38) specimen (water/binder series).

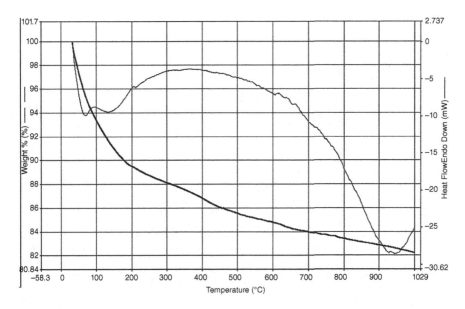

FIGURE 4.24 TGA/DTA curve for the C3 (water–binder ratio = 0.41) specimen (water/binder series).

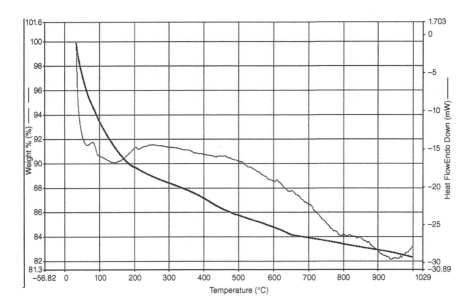

FIGURE 4.25 TGA/DTA curve for the C4 (water–binder ratio = 0.44) specimen (water/binder series).

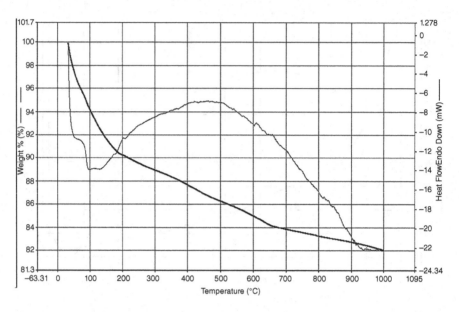

FIGURE 4.26   TGA/DTA curve for the C5 (water–binder ratio = 0.47) specimen (water/binder series).

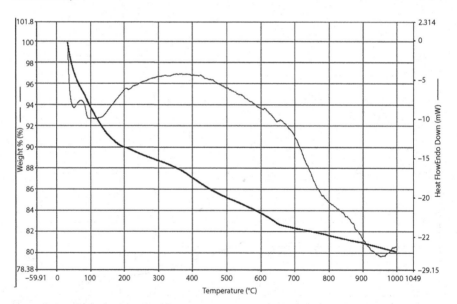

FIGURE 4.27   Water-binder ratio vs. weight retained.

weight retention of 83.19%. From the TGA graphs, it is observed that the water/binder ratio did not have a significant impact on weight retention percentage and it was in the range of 81–83% for all samples (Figure 4.27). Furthermore, the TGA graph followed the same trend for all specimens. An approximately 14% weight loss was noticed at 500°C.

4.3.1.8 Compressive Strength

4.3.1.8.1 Effect of Slag Content

The effect of slag on the compressive strength of AAC specimens has been reported in Figure 4.28. It is seen that the compressive strength decreases to a larger extent when exposed to elevated temperature with more slag content. The specimens with only fly ash as the source material recorded an increase in compressive strength with an increase in temperature up to 600°C and after 600°C, the strength decreased but lesser than the specimen with slag. The highest percentage decrease in strength was observed for specimen 40–60 and the lowest percentage decrease occurred for specimen 100-0.

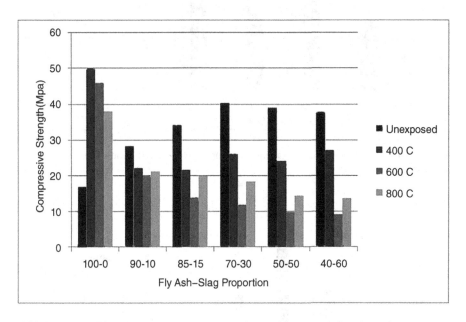

FIGURE 4.28 Effect of slag content on compressive strength. (exposed to elevated temperature)

4.3.1.8.2 Effect of Alkali Content

The effect of alkali content on compressive strength of AAC paste specimens at 400°C, 600°C, and 800°C is presented in Figures 4.29–4.31. The compressive strength of AAC specimens is dependent on the alkali content of the mix and exposure temperature. The maximum compressive strength of 30.43, 29.17, and 28 MPa was obtained for 8% alkali content at 400°C, 600°C, and 800°C, respectively. For 400°C, the minimum compressive strength recorded was 8.46 MPa for 4% Na_2O content. For 600°C, the minimum compressive strength of 7.86 MPa was obtained for the specimen with 4% Na_2O content. When the elevated temperature was increased to 800°C, the minimum compressive strength was recorded to be 7.43 MPa.

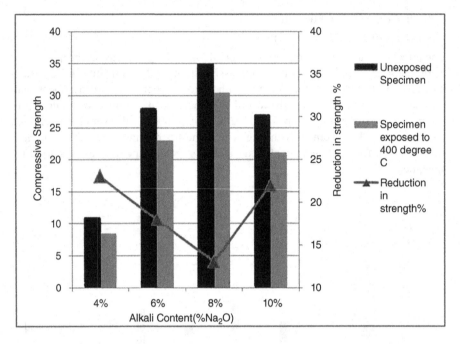

FIGURE 4.29 Effect of alkali content on compressive strength when exposed to 400°C temperature.

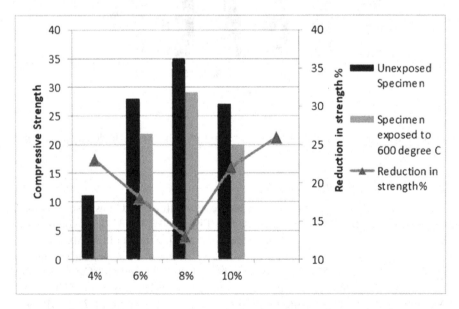

FIGURE 4.30 Effect of alkali content on compressive strength when exposed to 600°C temperature.

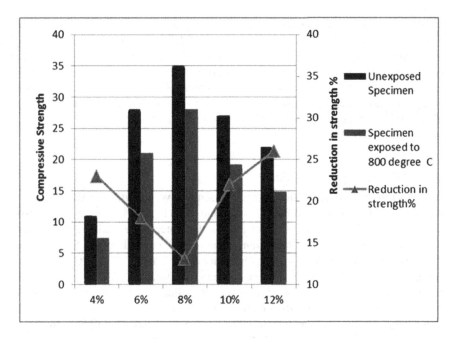

FIGURE 4.31 Effect of alkali content on compressive strength when exposed to 800°C temperature.

4.3.1.8.3 Effect of Silica Content

The effect of silica content on compressive strength of AAC paste specimens at 400°C, 600°C, and 800°C is presented in Figure 4.32. The compressive strength of AAC specimens is dependent on the silica content of the mix and exposure

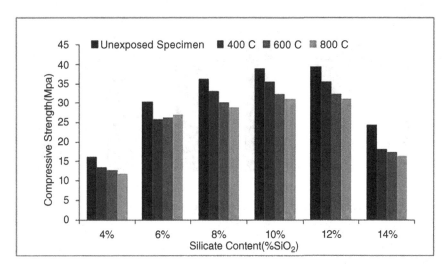

FIGURE 4.32 Effect of silica content on compressive strength when exposed to elevated temperature.

temperature. The maximum compressive strength was 35.59, 32.44 and 31.14 MPa for 12% silica content at 400°C, 600°C, and 800°C, respectively. For 400°C, the minimum compressive strength recorded was 13.56 MPa for 4% Na$_2$O content. For 600°C, the minimum compressive strength of 12.71 MPa was obtained for the specimen with 4% Na$_2$O content. When the elevated temperature was increased to 800°C, the minimum compressive strength was 11.79 MPa for the specimen with 4% Na$_2$O, respectively. Generally, strengths were seen to decrease with an increase in temperature.

4.3.1.8.4 Effect of Water Content

The effect of water/binder ratio content on compressive strength of AAC paste specimens at 400°C, 600°C, and 800°C is presented in Figure 4.33. The compressive strength of AAC specimens is dependent on the water/binder ratio of the mix and exposure temperature. The maximum compressive strength of 31.5, 28.71 and 27.60 MPa was obtained for 0.38 water binder content at 400°C, 600°C, and 800°C, respectively. For 400°C, the minimum compressive strength recorded was 17.2 MPa for a water/binder ratio of 0.47. For 600°C, the minimum compressive strength of 16.52 MPa was obtained for the sample with 0.47 water/binder ratio. When the elevated temperature was increased to 800°C, the compressive strengths were 23.46, 27.60, 21.58, 19.44, and 16.07 MPa for specimens C1-0.35, C2-0.38, C3-0.41, C4-0.44, and C5-0.47, respectively. Generally, strengths were seen to decrease with an increase in temperature and a higher water/binder ratio beyond an optimum point of 0.38 led to a decrease in strength.

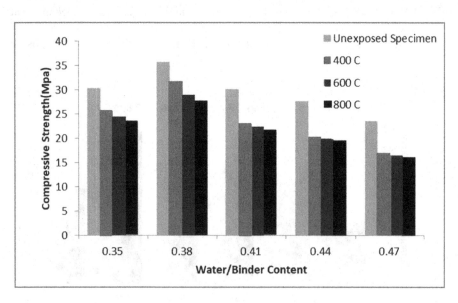

FIGURE 4.33 Effect of water/binder ratio on compressive strength of AACs when exposed to elevated temperature.

4.3.1.9 Microstructure Study by SEM/EDAX

4.3.1.9.1 Effect of Slag

It is seen clearly from Figure 4.34 that the specimen with the 10% slag shows a relatively more cohesive structure than the specimen with 15% and 30% slag, respectively. The mix 85-15, on the other hand, shows the presence of numerous small pores and thin cracks in the structure. The mix 70-30 shows the presence of much larger pores and cracks. It can be said that the addition of slag lessens the fire resistance of the AACs. It is noticed from Figure 4.35 that the calcium content increases with the increase of slag content and silica content increases with the increase in fly ash content.

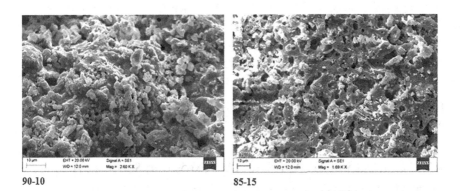

90-10   85-15

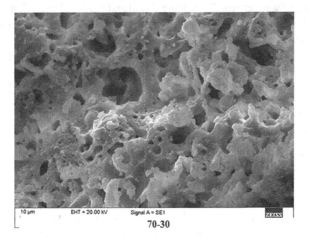

70-30

FIGURE 4.34 SEM images-effect of slag content on microstructure of AACs when exposed to elevated temperature.

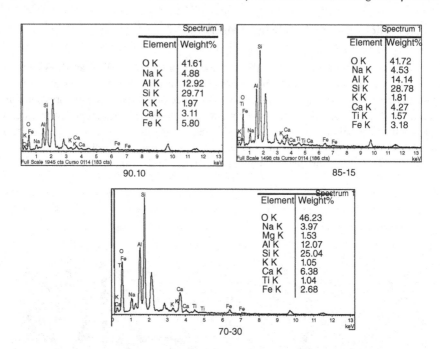

FIGURE 4.35 EDAX analysis-effect of slag content on composition of AACs when exposed to elevated temperature.

4.3.2 DURABILITY OF FLY ASH–SLAG AACs EXPOSED TO SULPHURIC ACID

4.3.2.1 Visual Appearance

The AAC specimens were exposed to sulphuric acid and remained intact. The surface did not display signs of severe deterioration but white and yellow colored precipitates were found to be deposited on the surface. The white colored precipitate increased in quantity with the increase in slag content in the mix. This precipitate is calcium sulphate, a result of the reaction between sulphate ions and the calcium in the slag. Thus, the amount of white precipitate in the form of calcium increases with the increase of slag content (Figure 4.36).

4.3.2.2 Weight Changes

4.3.2.2.1 Effect of Slag Content

Figure 4.37 indicates that the weight change decreases with the increase in slag content. In this investigation, slag was replaced only up to a percentage of 30% as that was the considered optimum percentage with regard to mechanical properties as observed in previous findings. The interesting fact about the findings is that the slag is less durable to sulphuric acid than fly ash but here addition of slag up to a certain percentage improves the resistance to sulphuric acid attack because of the dense pore structure. The addition of slag improves the reactivity level of the polymer system which leads to the formation of a tetrahedral three-dimensional framework to the higher extent. Thus, it is seen that the highest weight change is recorded by the mix 90-10 at 1.88% and the lowest weight change is recorded by mix 70-30 at 0.88%.

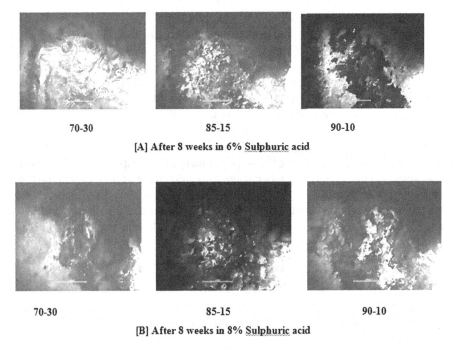

[A] After 8 weeks in 6% Sulphuric acid

FIGURE 4.36 AAC specimens exposed to 6% and 8% sulphuric acid solution for 8 weeks. (specimens having different fly ash–slag ratios)

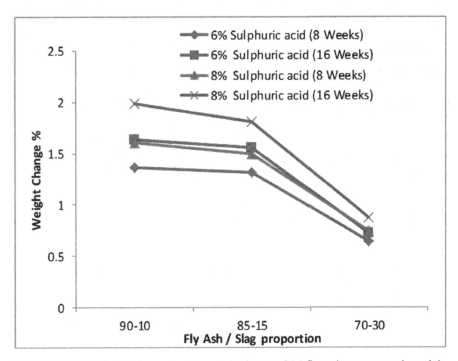

FIGURE 4.37 Effect of slag content on weight change of AAC specimens exposed to sulphuric acid solution.

4.3.2.2.2 Effect of Alkali Content, Silica Content, and Water/Binder Ratio

It is seen from Figures 4.38–4.40 that the percentage of synthesis parameters greatly affects the acid resistivity of the AACs. Here, it is seen that there is an increase in weight of the samples after it has been immersed in sulphuric acid solution of various concentrations for 8 weeks. The increase in alkali concentration corresponded with a decrease in the percentage of weight change with the A3-8%Na_2O specimen recording the lowest weight change percentage of 0.08%. This trend remained the same for the specimens which were dissolved for a longer duration of 16 weeks. The specimens immersed for 16 weeks also recorded an increase in weight with the A3-8% Na_2O sample recording 0.1% weight increase. The specimens also showed a decrease of percentage of weight change with an increase of SiO_2 content up to an optimum percentage of 12%. The water/binder ratio also played an important role in change in weight percentage of a specimen. The sample C2 with a water/binder ratio of 0.38 had the lowest weight change of 1.35%. The abovementioned findings

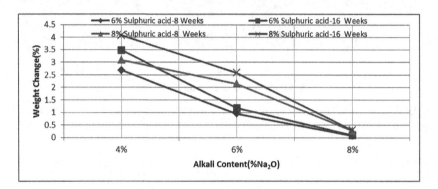

FIGURE 4.38 Effect of alkali content (%Na_2O) on weight change of AAC specimens exposed to sulphuric acid solution.

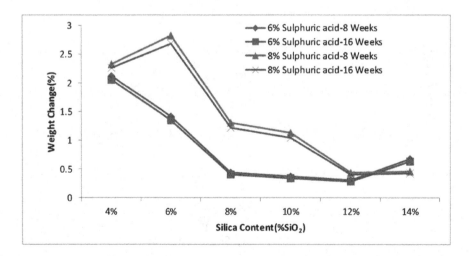

FIGURE 4.39 Effect of silica content (%SiO_2) on weight change of AAC specimens exposed to sulphuric acid solution.

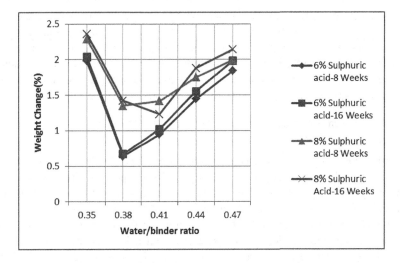

FIGURE 4.40   Effect of water/binder ratio on weight change of AAC specimens exposed to sulphuric acid solution.

point towards an interesting fact that the actual parameter governing the acid resistivity of the specimens is its porosity and tortuosity. All the best performing specimens, for example, A3, B5, and C2, exhibited a superior, compact, and denser microstructure with respect to the other specimens. An extended discussion on this aspect has been carried out in the following section where a relationship between sorptivity, apparent porosity, and acid resistance has been presented.

4.3.2.2.3 Relationship Between Weight Change and Water Sorptivity

A relationship between water sorptivity and weight loss of fly ash-based AAC specimens exposed to 6% sulphuric acid is shown in Figure 4.41. It is observed that the weight of the

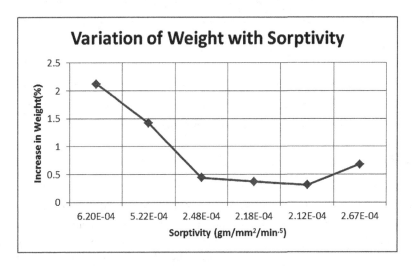

FIGURE 4.41   Relationship between water sorptivity and weight change of AAC specimens exposed to 6% sulphuric acid solution for 8 weeks.

specimen increases with the increase in sorptivity. The specimen possessing greater water sorptivity signifies that it has an extensive capillary network of interconnected voids which creates a pathway for the acid solution to penetrate into the specimen and attack the interface. This leads to the reaction between the polymerization products especially the ones having calcium as their dominant element with the sulphuric acid solution leading to de-alkalization and formation of additional reaction products which fill up the voids.

4.3.2.3 Residual Compressive Strength

4.3.2.3.1 Effect of Slag Content

Referring to Figure 4.42, it can be said that the highest residual compressive strength of 84% is found for the specimen out of mix 70-30 and the lowest residual compressive strength of 78% is recorded for the specimen out of the mix 90-10 when immersed in 6% H_2SO_4 solution for eight weeks. The same trend was observed with 6% H_2SO_4 specimens exposed to H_2SO_4 solution for 16 weeks. The mix 70-30 also recorded the highest residual compressive strength values when immersed in 8% H_2SO_4 for 8 and 16 weeks, whereas specimens out of mix 90-10 recorded the lowest values of residual compressive strength.

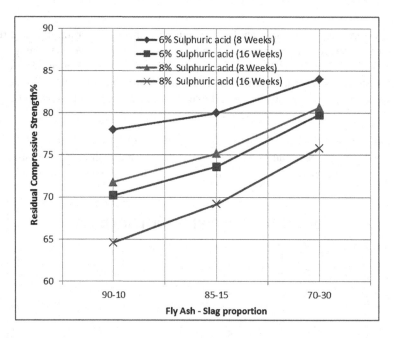

FIGURE 4.42 Relationship between residual compressive strength and slag content for AAC specimens exposed to sulphuric acid solution.

4.3.2.3.2 Effect of Alkali Content, Silica Content, and Water/Binder Ratio

Residual compressive strength is an important parameter while evaluating the durability of a material. Before conducting the compressive strength test, the specimens were kept for two days at room temperature for drying. The effect of residual alkali, silica, and water/binder ratio on residual strength is given in Figures 4.43–4.45, respectively. It is seen that increasing the Na_2O content up to an optimum content of 8% leads to an increase

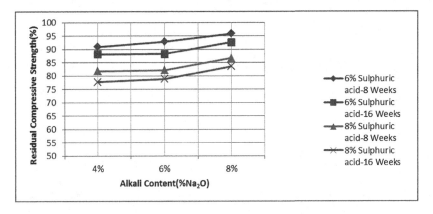

FIGURE 4.43 Effect of Na$_2$O content on residual compressive strength of AAC specimens exposed to sulphuric acid solution.

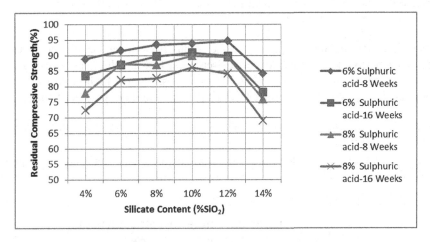

FIGURE 4.44 Effect of SiO$_2$ content on residual compressive strength of AAC paste specimens exposed to sulphuric acid solution.

in compressive strength due to a robust three-dimensional tetrahedral microstructure. During exposure to acid solutions, the specimens which had higher compressive strength before the exposure also recorded the highest residual compressive strength. The higher loss of strength in specimens having lesser Na$_2$O content had comparatively lesser residual compressive strengths than the other specimens. This phenomenon occurs due to the incomplete dissolution of the fly ash slag particles which leads to higher porosity and a less rigid gel structure. The specimens with a higher amount of SiO$_2$ up to a percentage of 12% recorded a higher residual compressive strength. Addition of SiO$_2$ up to an optimum amount leads to an increase in the intensity of the polymerization process, with the dissolvable silica serving as additional nucleation sites which leads to a better dissolution and higher degree of polycondensation which helps in building a gel network with less number of pores making it more resistant to acid attack. The water/binder ratio also governs the resistance of the AAC against sulphuric acid. In polymerization, water acts as a

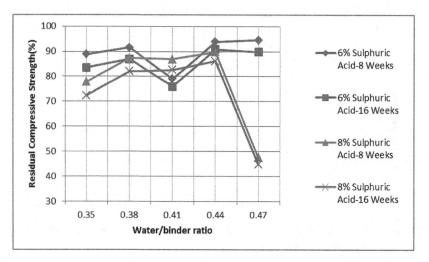

FIGURE 4.45   Effect of water–binder ratio on residual compressive strength of AAC paste specimens exposed to sulphuric acid solution.

transportation medium for the participating ions; thus excess water does not participate in the reaction process and evaporates during oven curing leaving behind pores which accelerates the ingression of the acid ions into the specimens. Figure 4.45 shows that the water binder ratio of 0.38 is the most suitable taking acid resistance into consideration.

4.3.2.3.3 Relationship between Residual Compressive Strength and Water Sorptivity

Figure 4.46 indicates that the lower the sorptivity, the higher the residual compressive strength. Specimens having higher strength possess a denser polymer matrix

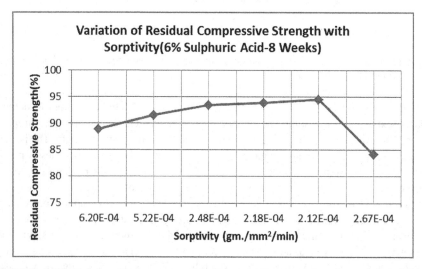

FIGURE 4.46   Relationship between residual compressive strength and sorptivity of AAC specimens exposed to 6% sulphuric acid solution.

which signifies lesser sorptivity. If the pore network is less extensive in nature, the acid ions are withheld from travelling deep into the gel network of the polymer. When the acid ions come in contact with the polymeric gel network, deleterious materials are formed which break the tetrahedral nature of the gel.

4.3.2.4 Microstructure Study by SEM/EDAX

The SEM images reiterate the fact that the microstructure of the AACs becomes more viscous and dense with the addition of slag (Figures 4.47–4.50). The voids in the microstructure also appear to lessen with the addition of slag. Now the lesser number of voids will resist the sulphate ions and thus the EDAX analysis based on the sample taken from surface exhibits a higher amount of sulfur percentage. The specimens which have been indicated to have higher strength had higher amount of sulfur detected during EDAX analysis, when the sample was taken from the surface of the test specimen. It may be said that as the acid ions are unable to travel inwards into the specimens, they get deposited on the surface.

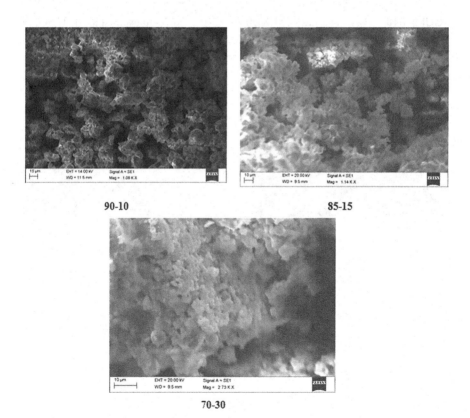

FIGURE 4.47   SEM images of MIX 90-10, 85-15, and 70-30 exposed to 6% H_2SO_4 for eight weeks (for different fly ash/slag ratios).

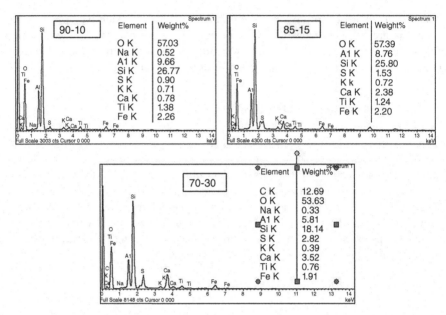

Element	Weight%
O K	57.03
Na K	0.52
Al K	9.66
Si K	26.77
S K	0.90
K K	0.71
Ca K	0.78
Ti K	1.38
Fe K	2.26

90-10

Element	Weight%
O K	57.39
Al K	8.76
Si K	25.80
S K	1.53
K k	0.72
Ca K	2.38
Ti K	1.24
Fe K	2.20

85-15

Element	Weight%
C K	12.69
O K	53.63
Na K	0.33
Al K	5.81
Si K	18.14
S K	2.82
K K	0.39
Ca K	3.52
Ti K	0.76
Fe K	1.91

70-30

FIGURE 4.48 EDAX analysis of MIX 90-10, 85-15, and 70-30 exposed to 6% H_2SO_4 for eight weeks. (for different fly ash/slag ratios)

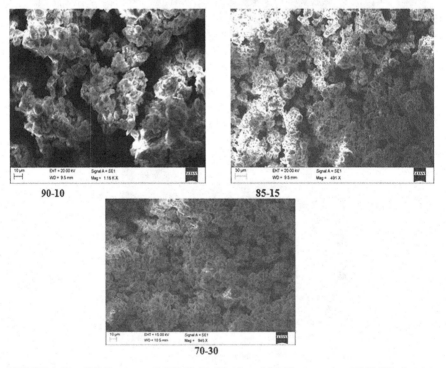

90-10

85-15

70-30

FIGURE 4.49 SEM images of MIX 90-10, 85-15, and 70-30 exposed to 8% H_2SO_4 for eight weeks. (for different fly ash/slag ratios)

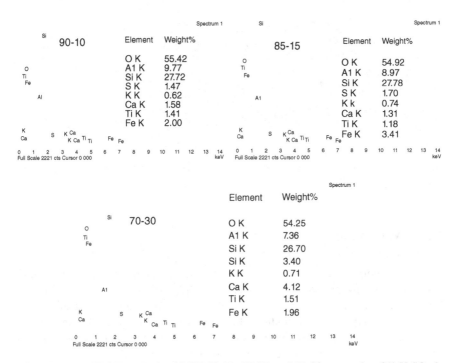

FIGURE 4.50   EDAX analysis of MIX 90-10, 85-15, and 70-30 exposed to 8% H_2SO_4 for eight weeks. (for different fly ash/slag ratios)

4.3.3 Durability of Fly Ash–Slag AACs Exposed to Nitric Acid Solution

4.3.3.1 Visual Appearance

In general, no visible changes in the shape and size of AAC specimens were observed when exposed to nitric acid solution. Disintegration of specimen does not occur, though some corrosion of exposed surfaces is visible. However, slight changes in color are noticed which turns light yellow from the initial light grey. Surface appearances of AAC paste specimens having different Na_2O contents after 8 weeks and 10 weeks of exposure to 6% and 8% nitric acid solution are shown in Figures 4.51 and 4.52. It is evident that AAC specimens with lesser slag content have undergone more surface corrosion. Moreover, the surface corrosion increased with time.

FIGURE 4.51   AAC specimens after eight-week exposure to 6% nitric acid solution.

90-10   85-15   70-30

FIGURE 4.52 AAC specimens after 8eight-weeks exposure to 8% nitric acid solution. (for different fly ash/slag ratios)

4.3.3.2 Weight Change

4.3.3.2.1 Effect of Slag Content

Figure 4.53 indicates that the negative weight change (%) decreases with the increase in slag content. In this investigation, slag was replaced only up to a percentage of 30% as that was the considered optimum percentage with regard to mechanical properties as observed in previous findings. In comparison to the weight gain in sulphuric acid, there is a weight loss of the specimens exposed to nitric acid. It is seen that the rate of decrease of weight lessens with addition of slag content as it leads to the

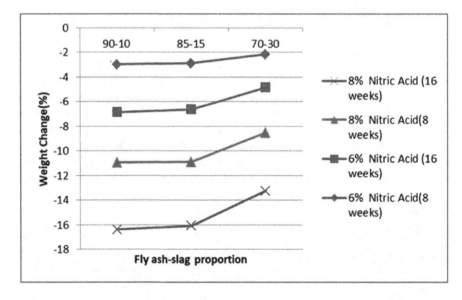

FIGURE 4.53 Relationship between weight change and slag content of AAC specimen exposed to nitric acid.

formation of a robust three-dimensional gel structure. Thus, it is seen that the highest weight change is recorded by the mix 85-15 at 5.20% and the lowest weight change is recorded by mix 70-30 at 2.16%.

4.3.3.2.2 Effect of Alkali Content

As mentioned in the earlier section, Na_2O plays an important role with regard to acid resistivity of an AAC. As opposed to its behavior in sulphuric acid where there was an increase in weight after immersion in acid, the AACs, however, undergo mass loss when immersed in nitric acid. This percentage of mass change is observed to increase with lower Na_2O content. Higher Na_2O contents lead to an increased rate of dissolution, which leads to a greater amount of polymeric gel formation. When the gel formation increases, there is a decrease in the porosity of the specimen thus blocking the path of the deleterious acid ions. However, the Na_2O has to be added up to an optimum amount, for example, 8% in this case. Excess Na_2O leads to accelerated dissolution leading to flash setting. Referring to Figure 4.54, it is seen that specimens with 4% Na_2O record a negative weight change percentage of 1.93 and the sample with 8% Na_2O records a negative weight change percentage of 0.84.

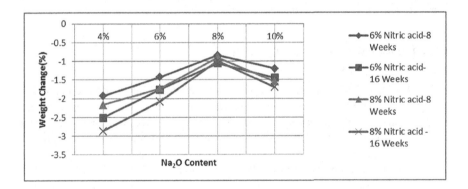

FIGURE 4.54 Relationship between weight change and Na_2O content (%) of AAC exposed to nitric acid solution.

4.3.3.2.3 Relationship of Weight Change and Apparent Porosity

Figure 4.55 shows a relationship between weight change and apparent porosity of a specimen. The specimen having the lowest apparent porosity of 5% has a decrease in weight of 0.84% but the sample with highest apparent porosity of 13% has the least weight loss of 1.93%.

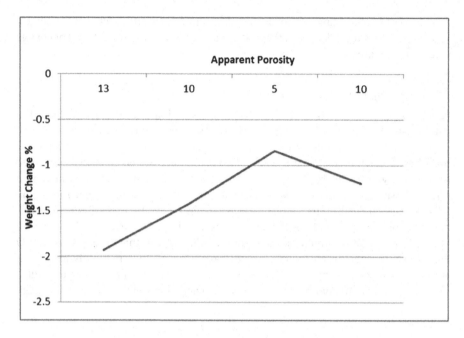

FIGURE 4.55 Relationship between weight change and apparent porosity of AAC specimens exposed to 8% nitric acid solution.

4.3.3.3 Residual Compressive Strength

4.3.3.3.1 Effect of Slag Content

Figure 4.56 shows a relationship between slag content and residual compressive strength of different AAC specimens. The residual compressive strength values are higher for the specimens having a higher slag content. Here, it is seen that for the samples immersed for 8 and 16 weeks, the lowest residual compressive strength values are 78% and 71.76%, respectively, for mix 90-10 immersed in 6% HNO_3 solution. The mix 70-30 immersed in 6% HNO_3 solution recorded higher values of 84% and 80% of residual compressive strength for 8 weeks and 16 weeks of exposure, respectively. The trend is similar for specimens immersed in 8% HNO_3 solution for 8 weeks as well as 16 weeks.

4.3.3.3.2 Effect of Alkali Content

The residual compressive strength of the specimens exposed to acid basically depends upon the penetration of the acid ions into the polymer gel network. As the amount of Na_2O is increased up to an optimum extent, the polymer gel network becomes more stable as sodium plays the role of a charge balancer. It is seen from Figure 4.57 that the specimen with Na_2O of 4% recorded a residual compressive strength of 81.22% and the sample with 8% Na_2O recorded a residual compressive strength of 93%. However, the lowest residual compressive strength was recorded with a Na_2O percentage of 10% as increasing the Na_2O content above an optimum amount results in incomplete polymerization leading to lesser strength.

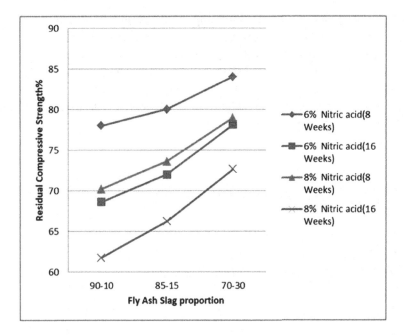

FIGURE 4.56   Effect of slag content on residual compressive strength of AAC specimens exposed to nitric acid solution.

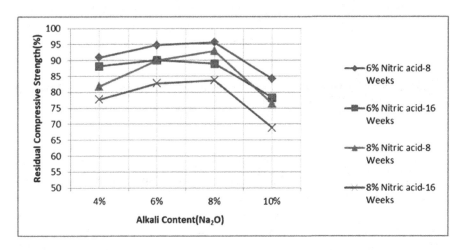

FIGURE 4.57   Residual compressive strength of AAC specimens exposed to nitric acid solution.

4.3.3.3.3 Relationship of Residual Compressive Strength and Apparent Porosity of AACs Exposed to 8% Nitric Acid Solution

Figure 4.58 shows that the residual compressive strength of the AAC specimens increases with the decrease in apparent porosity. The optimum dosage (8%) of the alkaline activator solution causes the formation of a dense, robust microstructure with a lesser number of voids which leads to an increase in mechanical strength as well as acid resistance of the specimens.

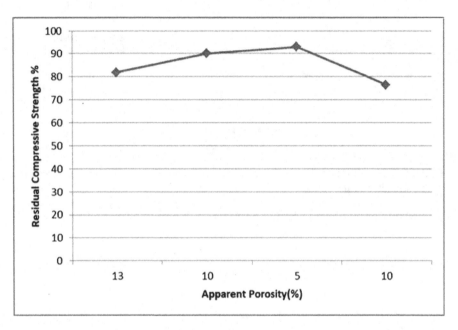

FIGURE 4.58   Relationship between residual compressive strength and apparent porosity of AAC exposed to 8% nitric acid solution.

4.3.3.3.4 Microstructure Study by SEM/EDAX

The SEM images and EDAX analysis shown in Figures 4.59 and 4.60 corroborate the fact that the 70-30 mix specimen possesses better resistivity against nitric acid. The gel structure of the 70-30 mix appears to be the most compact and devoid of voids. The images also prove that the porosity is the determining factor regarding the acid resistivity of the specimens.

90-10   85-15

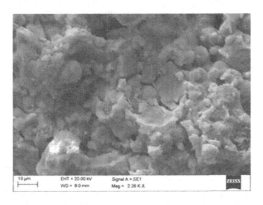

70-30

FIGURE 4.59 SEM images of MIX 90-10, 85-15, and 70-30 exposed to 8% HNO₃. (for different fly ash/slag ratios)

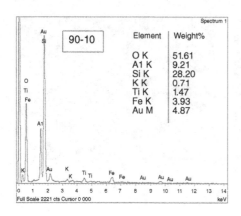

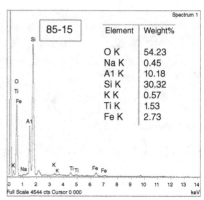

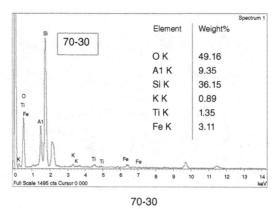

70-30

FIGURE 4.60 EDAX analysis of mix 90-10, 85-15, and 70-30 exposed to 8% HNO_3. (for different fly ash/slag ratios)

4.3.4 DURABILITY OF FLY ASH-BASED AACS EXPOSED TO MAGNESIUM SULPHATE SOLUTION

4.3.4.1 Visual Appearance

Fly ash–slag-based AAC specimens were exposed to 6 and 8 percent magnesium sulphate solution for 8 and 16 weeks, respectively. Slag content, alkali content, silica content, and water/binder ratio were varied. The surface of the specimens was examined closely using an optical microscope and granule-like white deposits were observed to be deposited on most of them. The granules at first were semi solid in nature but with time solidified to a great extent. A few selected photographs of the various specimens have been provided above. On visually inspecting the specimens with varying alkali content, it is seen from Figures 4.61 and 4.62 that the amount of precipitate being deposited increased with the time of exposure. The deposits when subjected to EDAX analysis reported the presence of a sodium-alumino-sulphate compound and a calcium alumino-sulphate under prolonged exposure in magnesium sulphate solution. These deposits were mostly found at the surface of the specimen.

FIGURE 4.61   AAC specimens exposed to 8% magnesium sulphate solution for eight weeks.

FIGURE 4.62   AAC specimens after exposure to 8% magnesium sulphate solution

On crushing these specimens, the inner structure was found to bear resemblance to the unexposed specimens.

4.3.4.2 Weight Change

4.3.4.2.1 Effect of Alkali Content (%Na$_2$O)

Prior to exposition in the magnesium sulphate solution, the specimens were immersed in water for three days and were then subsequently taken out and its weight was measured. After this procedure was completed, the specimens were submerged in magnesium sulphate solutions for 8 and 16 weeks. At the end of the respective duration of exposure, the specimens were taken out of the magnesium sulphate solutions and the surface precipitates were scraped and cleaned and then weight is taken. Samples A1, A2, A3, A4, and A5 record an increase in weight of 0.57, 0.41, 0.97, 0.45, and 0.64%, respectively. From Figures 4.63 and 4.64, it is seen that there is a decrease in weight change values with the increase in Na$_2$O content up to an optimum content of 8%. This trend is in sync with the results of mechanical properties of the unexposed fly ash–slag specimens which had also shown an increase up to an optimum value of 8% Na$_2$O. This clearly indicates that up to a certain value of Na$_2$O percentage there is an improvement

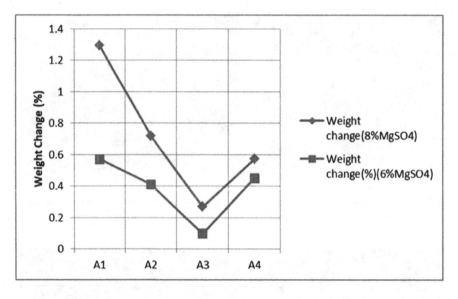

FIGURE 4.63 Effect of Na$_2$O content on weight change of AAC specimens exposed to 6% and 8% magnesium sulphate solution.

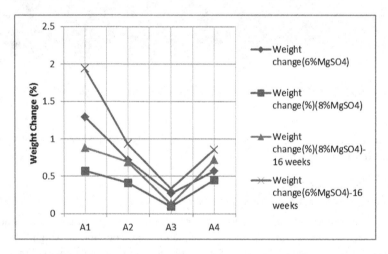

FIGURE 4.64 Effect of Na$_2$O content on weight change of AAC specimens exposed to 6% and 8% magnesium sulphate solution for a duration of 8 and 16 weeks.

in the microstructure of the AAC specimens which aids in lessening the intensity of creation of sulphate-based reaction products. It is also seen that with the increase in the concentration of the solution to 8%, changes in weight of each of the specimens followed the same trend as reported when exposed to 6% MgSO$_4$ concentration. Specimen A1 reported the highest weight increase of 1.29% over A2, A3, A4, and A5 reporting an increase of 0.72%, 0.27%, 0.57%, and 1.11%, respectively. The weight changes in the specimens when exposed to different periods of duration, for example, 8 weeks and 16 weeks, were also measured, and the values have been plotted in Figures 4.63 and 4.64 which clearly indicates that the specimen with the least amount of Na_2O, for example, A1 specimen recorded the highest increase in weight. The weight change values of the specimens of the alkali series clearly indicate that the Na_2O content present in the activator solution plays a major role in determining the behavior against sulphate exposure. It is already appreciated that the increase in the value of the Na_2O content up to an optimum percentage (8%) has an effect on the apparent porosity as well as sorptivity of the fly ash–slag AAC specimens, and these two factors play an important role in the determination of the level of permeation. This level of permeation to a great extent determines the formation of the new reaction products in the existing pores of the AAC specimens.

4.3.4.2.2 Effect of Silica Content

The results of the weight change of the samples with varying amount of SiO_2 percentage when subjected to exposure in 6% and 8% magnesium sulphate are presented in Figure 4.65. It is seen that there is a decrease in the weight change with the increase of SiO_2 up to an optimum percentage of 10%, for example, the sample labeled B4 recorded the lowest weight change value. The samples B1, B2, B3, B4, B5, and B6 reported a weight change of 1.87%, 1.73%, .89%, 0.23%, 0.73%, and 0.71% with 6% $MgSO_4$ and for 6% $MgSO_4$ are 2.17%, 1.99%, 0.733%, 0.278%, 0.733%, and 0.533%, respectively. Thus from the values it can be said that in the solutions of different concentrations, there is decrease in weight loss up to 10% SiO_2 (specimen B4) and then the weight loss percentage increased for specimens B5 and B6 having 12% and 14% SiO_2 content, respectively. It was seen that the B6 specimen which is exposed to 8% $MgSO_4$ recorded less weight loss than the B5 sample.

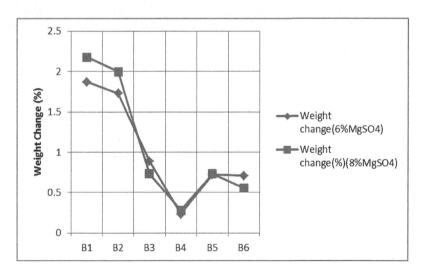

FIGURE 4.65 Effect of SiO_2 content on weight change of AAC specimens exposed to 6% and 8% magnesium sulphate solution.

The B6 sample exposed to 6% $MgSO_4$ recorded a weight loss similar to the B5 sample. Thus it clearly seen that the behavior of the AAC specimens in magnesium sulphate exposure is related to the SiO_2 content in the activator solution.

4.3.4.3 Relationship of Weight Gain and Apparent Porosity

The values of the respective weight changes of the samples of the alkali series and silica series were plotted against their respective apparent porosities. From Figures 4.66 and 4.67, it is said that the weight change of the specimens is related to the

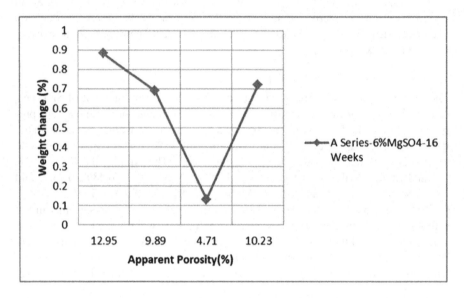

FIGURE 4.66 Relationship between weight gain and apparent porosity of AAC specimens with varying amount of silica exposed to 6% magnesium sulphate solution. (alkali series)

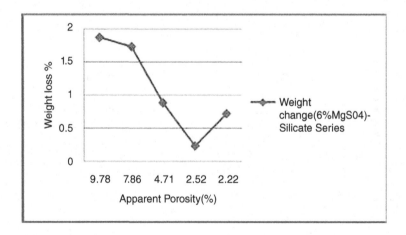

FIGURE 4.67 Relationship between weight gain and apparent porosity of AAC specimens with varying amount of silica exposed to 6% magnesium sulphate solution. (silicate series)

apparent porosities of the concerned specimens. It is seen in Figure 4.67 that the A1 specimen of porosity 12.95% has undergone a weight change of 0.88% with A2 and A3 samples recording 0.69% and 0.13% weight change at an apparent porosity of 9.89% and 4.71%, respectively, but for the A4 sample, the weight change increased to 0.72% at an apparent porosity of 10.23%. This relationship between weight change and apparent porosity can further be appreciated from the data recorded for the specimens belonging to the silica series. The specimen B1 has a weight change value of 1.89% against an apparent porosity of 9.78%, and the weight change values of the B2, B3, B4, and B5 samples are 1.72%, 0.88%, 0.23%, and 0.72%, respectively.

4.3.4.4 Residual Compressive Strength

4.3.4.4.1 Effect of Alkali Content

It has been seen in previous studies that the increase in the concentration of Na_2O content increases the rate of dissolution. With the increase in the rate of dissolution, the polymerization occurs at a faster rate and also improves the nature of the three-dimensional gel network formed in AACs. It has also been previously reported that the nature of the three-dimensional gel network depends upon the nature of the network modifying cations. In the fly ash slag system, the charge balancing cation is provided by the Na^+ ions of the activator solution and Ca^+ ions present in the slag. When the fly ash slag samples are immersed in the magnesium sulphate solution, the Mg+ ions enter into the three-dimensional polymeric gel network and lower its stability as well as rigidity. It can be seen from the results shown in Figure 4.68 that the A1 sample retained 75% of its compressive strength, A2, A3, and A4 had residual compressive strength values of 63.86%, 84.21%, and 62.14% and 48.46%, 78.05%, and 50.49% in 6% $MgSO_4$ and 8% $MgSO_4$ concentrations, respectively. From the SEM images (Figure 4.71) shown in the later section, it can be seen that the A3 had

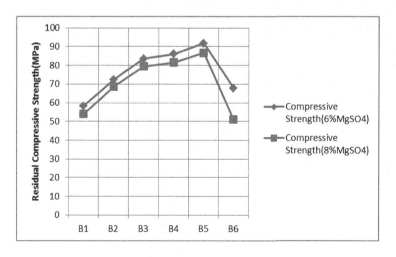

FIGURE 4.68 Effect of Na_2O content on residual compressive strength for AAC paste specimens exposed to 8% magnesium sulphate solution.

a more rigid and cohesive gel structure than the other samples and it has helped retain 84.21% and 78.05 in 6% and 8% MgSO₄, respectively Though A1 had a larger residual compressive strength value than A2, it had a more porous structure than it as is evident from the apparent porosity values as well as the SEM images (Figure 4.71). The residual compressive strength value was high as it had an extremely low actual compressive strength. The specimen A4 recorded the lowest value of 62.14% and 50.49% exposed to 6% and 8% MgSO₄ concentrations. This result also tallied with the fact that A4 had a greater apparent porosity value than the other samples.

4.3.4.4.2 Effect of Silica Content

The values of the residual compressive strength exposed to 6% and 8% $MgSO_4$, and compressive strengths have been plotted against their respective specimen numbers in Figure 4.69 The samples B1, B2, B3, and B4 exposed to 6% $MgSO_4$ and 8% $MgSO_4$ solution recorded a weight change percentage of 1.87%, 1.72%, 0.88%, and 0.233% and 2.17%, 1.99%, 0.733%, and 0.277%, respectively. Samples B5 and B6 exposed 6% $MgSO_4$ and 8% $MgSO_4$ recorded a weight change of 0.72%, 0.71% and 0.733%, 0.558% respectively. Figure 4.69 shows that with the increase up to an optimum percentage of SiO_2, there is an increase in residual compressive strength and beyond that, the values start to decrease. The addition of silica content up to an optimum percentage causes an increase in the intensity of condensation of the polymeric gel network thus leading to a more robust three-dimensional gel network. The deterioration caused by the migration of the Mg^+ ions and SO_4^- ions into the binder systems depends upon the porosity of the samples. If the pore network is more expansive in nature, then the Mg^+ ions will find a path for transportation into the internal

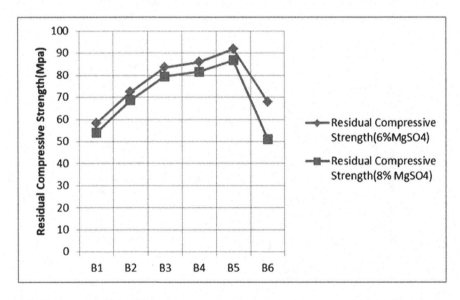

FIGURE 4.69 Effect of SiO_2 content on residual compressive strength of AAC specimens exposed to 6% and 8% magnesium sulphate solution.

structure of the specimens. Thus, the addition of silica up to a certain percentage can increase the resistivity to the penetration of Mg^+ ions as well as SO_4^- ions.

4.3.4.5 Relationship of Residual Compressive Strength and Apparent Porosity

A correlation has been tried to establish between apparent porosity of the AAC sample and residual compressive strength when exposed to 6% magnesium sulphate solution for 8 weeks. It can be seen from Figure 4.70 depicting the behavior of the specimens with varying amount of silica that the residual compressive strength value increases with decreasing amount of porosity. In the results of alkali series, it is seen that the residual compressive strength value at 9.78% porosity of the A1 sample is greater than its corresponding value at 7.81% porosity which is because of the fact that the A1 sample recorded an extremely low initial compressive strength. However, a look at the actual compressive strength values of the specimens exposed to magnesium sulphate solution will show that the increase in apparent porosity had a negative effect on the residual compressive strength.

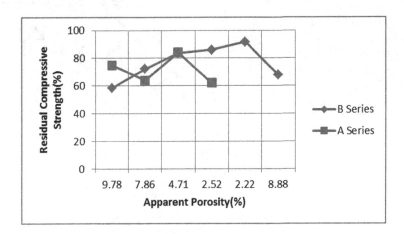

FIGURE 4.70 Relationship between residual compressive strength and apparent porosity of AACs exposed to 6% magnesium sulphate solution.

4.3.4.6 Microstructure Study by SEM/EDAX

The samples for the SEM analysis were collected from the surface of the specimens. On observing the SEM images shown in Figure 4.71, it may be said that certain precipitation products could be found in addition to the polymeric gel. The precipitate could be classified as a magnesium calcium alumino silicate phase. Thus, it can be seen that the magnesium ions have penetrated into the sodium alumino calcium sulphate gel. While comparing the specimens exposed to 8 weeks and 16 weeks of exposure, respectively, it is seen that the specimens exposed for a greater period of time had lesser voids, possibly due to the deposition of precipitation products. The residue was formed at the surface of the specimens, and magnesium, iron, and sulfur are the major constituents.

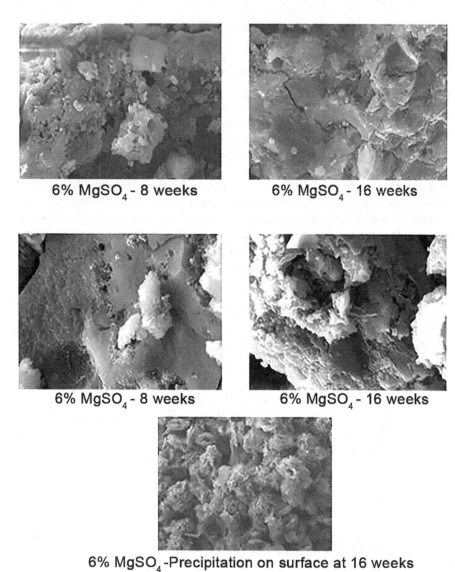

6% MgSO$_4$ - 8 weeks   6% MgSO$_4$ - 16 weeks

6% MgSO$_4$ - 8 weeks   6% MgSO$_4$ - 16 weeks

6% MgSO$_4$ -Precipitation on surface at 16 weeks

FIGURE 4.71   SEM images and EDAX data of the specimen (70-30) exposed to 8% MgSO$_4$ solution.

4.3.5 DURABILITY OF FLY ASH–SLAG AAC SPECIMENS EXPOSED TO SODIUM SULPHATE SOLUTION

4.3.5.1 Visual Appearance

Fly ash-based AAC specimens were exposed to 6 and 8% sodium sulphate solution for 8 and 16 weeks and are shown in Figures 4.72–4.75. Slag content, alkali content, silica content, and water/binder ratio were varied. The surface of the specimens was examined closely using an optical microscope and unlike the specimens subjected to magnesium sulphate solution, white granule-like deposits were scarce in nature. Specimens with high alkali and silica content did not have any visible deposits and resembled the unexposed AAC specimens. The extent of the deposits varied with changing parameters which will be further discussed in the following sections.

FIGURE 4.72 AAC specimens exposed to 6% sodium sulphate solution after 16 weeks.

FIGURE 4.73 AAC specimens exposed to 8% sodium sulphate solution after 16 weeks.

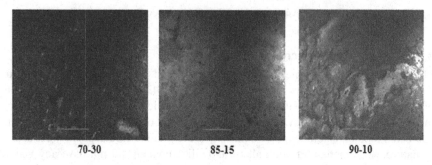

FIGURE 4.74   AAC specimens exposed to 6% sodium sulphate solution after eight weeks.

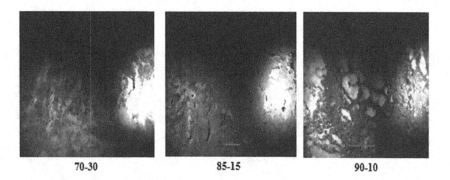

FIGURE 4.75   AAC specimens exposed to 8% sodium sulphate solution after eight weeks.

4.3.5.2 Weight Change

4.3.5.2.1 Effect of Alkali Content

Prior to exposing in the sodium sulphate solution, the specimens were immersed in water for three days and then taken out for measuring weight. Then the specimens were submerged in sodium sulphate solutions for 8 and 16 weeks. At the end of the respective duration of exposure, the specimens were taken out of the sodium sulphate solution, and the surface precipitates were scraped and cleaned and then weight was taken. Samples A1, A2, A3, and A4 recorded an increase in weight of 0.5, 0.37, 0.1, and 0.52 percent, respectively. As per Figure 4.76, it is seen that there is a decrease in weight change values with the increase in Na_2O content up to an optimum content of 8%. This behavior is similar to the one demonstrated by specimens immersed in magnesium sulphate solution. It is also seen that with the increase in concentration of the sodium sulphate solution to 8%, the weight change percentage continues to decrease up to 8% alkali (%Na_2O). A4 specimens with Na_2O content of 10% reported an increase in weight change percentage exposed to 6% and 8% sodium sulphate solution.

The weight changes in the specimens when exposed 8 weeks and 16 weeks were also measured. A3 reporting the lowest change in weight change percentage, whereas A1 and A4 reported comparatively higher values. The weight change value decreases with increasing content of Na_2O up to an optimum percentage of 8%.

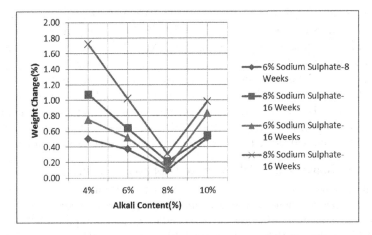

FIGURE 4.76 Effect of Na$_2$O content on weight change of AAC specimens exposed to sodium sulphate solution.

4.3.5.2.2 Effect of Silica Content

The results of the weight change of the samples with varying amount of SiO$_2$ percentage when exposed to 6% and 8% sodium sulphate are presented in Figure 4.77. It is seen that there is a decrease in the weight change with the increase of SiO$_2$ up to an optimum percentage of 10%, for example, the specimen B4 recorded the lowest weight change value. The samples B1, B2, B3, B4, B5, and B6 reported a weight change of 0.52%, 0.49%, 0.27%, 0.16%, 0.12%, and 0.52% when exposed to 6% Na$_2$SO$_4$ are 0.78%, 0.71%, 0.52%, 0.23%, 0.18%, and 0.74%, respectively. The trend of the results mirrors the behavior of the specimens submerged in magnesium

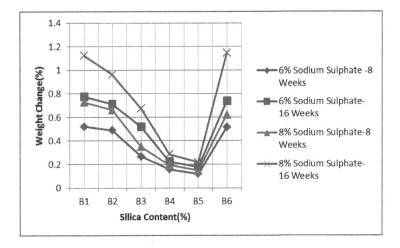

FIGURE 4.77 Effect of silica content (SiO$_2$) on weight change for AAC specimens exposed to sodium sulphate solution.

sulphate solution. The trend also remains unchanged when the specimens are exposed to, for example, 8 and 16 weeks. The weight change percentage of each of the specimens increases when exposed to 8% Na_2SO_4 solution with respect to the specimens immersed in the 6% Na_2SO_4. It may be said that the ingress of sulphate ions depends on the microstructure of the specimens. The microstructure of AAC specimens depends partly on the silicate concentration which controls the rate of polymerization. The increase in silicate concentration up to an optimum content of 10% leads to a greater intensity of polycondensation which lessens the percentage of porosity in the AAC specimen which in turn prevents the entry of sulphate ions.

4.3.5.3 Relationship of Weight Gain and Apparent Porosity

The values of the respective weight changes of the specimens of the alkali series were plotted against their respective apparent porosities. From Figure 4.78, it is seen that the weight change of the sample is related to the apparent porosity. It is seen in Figure 4.3.5.3 that the B1(4%SiO_2) specimen of porosity 9.78% has undergone a weight change of 1.12%, whereas B2 and B3 specimens has recorded 0.96% and 0.67% weight change at an apparent porosity of 7.86% and 4.71%, respectively. The rate of weight change continued to decrease with the increase in silica content for B4 and B5 specimens but for the B6 specimen, the weight change increased to 1.15% at an apparent porosity of 8.88%.

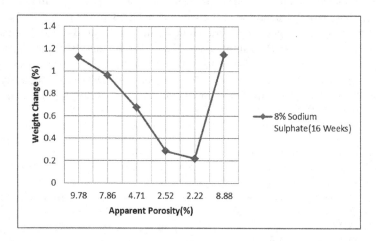

FIGURE 4.78 Relationship between weight gain and apparent porosity of AAC specimens with varying amount of silica exposed to 6% and 8% sodium sulphate solution.

4.3.5.4 Residual Compressive Strength

4.3.5.4.1 Effect of Alkali Content

It is seen from the results shown in Figure 4.79 that the A1(4%Na_2O), A2(6%Na_2O), A3(8%Na_2O), and A4(10%Na_2O) had residual compressive strength values of 72.32%, 80.61%, 92.11%, and 76.44% when exposed to 6% Na_2SO_4 and 71.55%, 73.78%, 90.13%, and 66.66% when exposed to 8% Na_2SO_4,

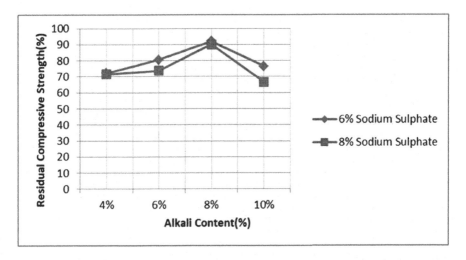

FIGURE 4.79   Effect of Na_2O content on residual compressive strength for AAC specimens exposed to 6% and 8% sodium sulphate solution.

respectively. The residual compressive strength of the specimens is seen to increase with the increase of silicate content. In the case of polymeric specimens, the compressive strength is linked with the rigidity of the polymeric gel structure. The nature of the gel network depends greatly upon the rate of dissolution of the source material. Higher alkali concentration leads to the increase in the intensity of dissolution resulting in the increase of rate of leaching of Si^{4+} and Al^{3+} ions. Additional silica and aluminum ions promote the formation of comparatively more rigid polymeric tetrahedral network. As the rigidity of the polymeric gel network increases so does its resistivity towards ingress of sulphate ions. The lesser the amount of sulphate ions able to penetrate the polymeric gel network and break the tetrahedral framework, the greater is the value of the residual compressive strength. The results also demonstrate that the percentage of residual strength preserved is much higher for specimens immersed in magnesium sulphate. The reason for a greater level of deterioration of the specimens exposed to $MgSO_4$ solution can be attributed to the fact of the replacement of the Na+ cation in the polymeric framework with the Mg+ ion. This cation exchange causes instability in the gel network leading to the lessening of compressive strength. In the case of sodium sulphate solution, the cation of the sulphate solution and the activator solution is the same; thus the interchange of cations does not harm the stability of the gel network.

4.3.5.4.2 Effect of Silica Content

The values of the residual compressive strength under the exposure to 6% and 8% Na_2SO_4 have been plotted in Figure 4.80. The trends are of the same nature of both the graphs in Figure 4.80. The specimens B1(4%SiO_2), B2(6%SiO_2), B3(8%SiO_2), B4(10%SiO_2), B5(12%SiO_2), and B6(14%SiO_2) recorded a residual compressive strength of 66.67%, 80.28%, 92.47%, 92.88%, 93.25, and 82.79%

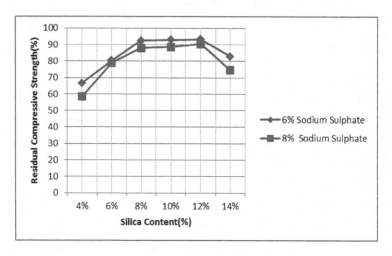

FIGURE 4.80   Effect of SiO$_2$ content on residual compressive strength of AAC specimens in 6% and 8% sodium sulphate solution for 16 weeks.

when exposed to 6% Na$_2$SO$_4$ and 58.33%, 78.87%, 87.9%, 87.61%, 90.69%, and 74.59% when exposed to 8% Na$_2$SO$_4$ solution respectively. The increase of silicate content increases the rate of polycondensation which in turn leads to the generation of a robust three-dimensional gel network which is observed to be similar to the results of the residual compressive strengths of the specimens immersed in the sodium sulphate solutions, and is higher when compared with the specimens exposed to magnesium sulphate solutions. This is due to the nature of the gel structure which becomes disrupted due to the ingress of magnesium ions. However in the case of sodium sulphate, the cations of the sulphate solutions participate in the polycondensation process and take part in the formation of a sodium alumino-calcium silicate gel.

4.3.5.5 Microstructure Study by SEM/EDAX

On observation of the SEM images shown in Figures 4.81 and 4.82, it is seen that as the alkali content increases, the microstructure of the fly ash-based AAC samples becomes denser. The improved microstructure contains fewer pores which prevents the ingress of the sodium sulphate ions. This argument can be further supported by the EDAX analysis shown in Figures 4.83 and 4.84, where it is seen that the sulphur content is reducing with the increase in the alkali content up to the optimum percentage of 8% Na$_2$O. It is also seen that when the concentration of the sodium sulphate solution is increased from 4% to 6%, the sulfur content increases in all specimens irrespective of the alkali content when exposed to 4% Na$_2$SO$_4$. The SEM and EDAX indicate the fact that the porosity of the specimens is a governing factor in the sulphate-resistant properties of the fly ash-based AAC.

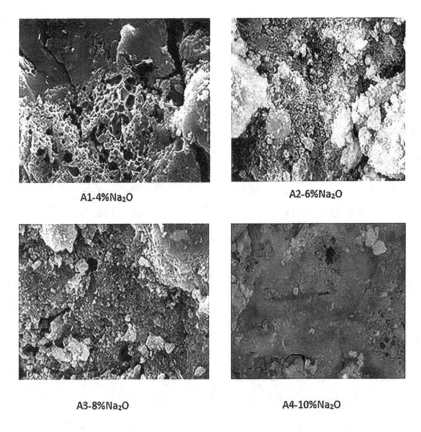

A1-4%Na₂O

A2-6%Na₂O

A3-8%Na₂O

A4-10%Na₂O

FIGURE 4.81 SEM images of AAC specimens AAC specimens AAC specimens exposed to 6% sodium sulphate solution after 16 weeks.

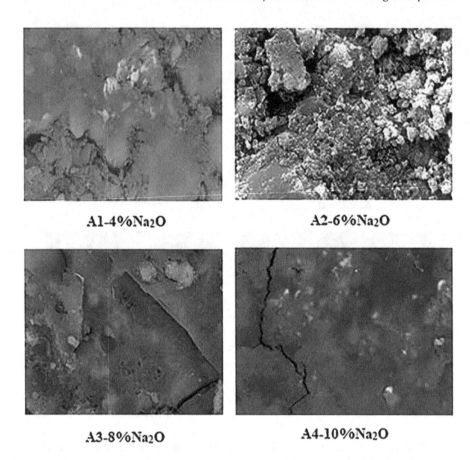

A1-4%Na₂O   A2-6%Na₂O

A3-8%Na₂O   A4-10%Na₂O

FIGURE 4.82   SEM images of AAC specimens exposed to 8% Na$_2$SO$_4$ for 16 weeks.

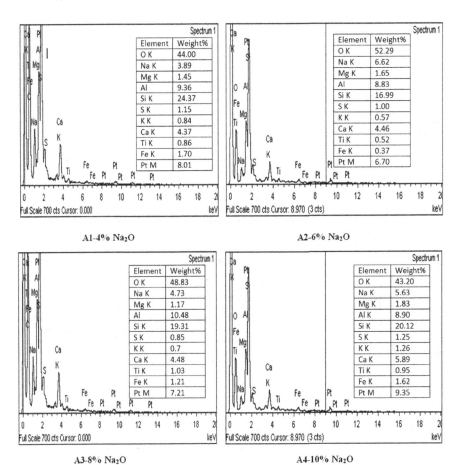

FIGURE 4.83 EDAX analysis of AAC specimens exposed to 6% Na_2SO_4 for 16 weeks.

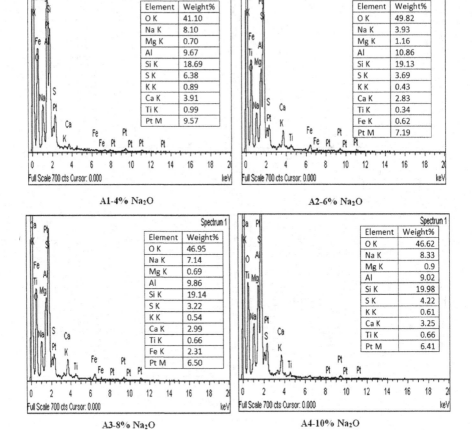

FIGURE 4.84 EDAX analysis of AAC specimens exposed to 8% Na_2SO_4 for 16 weeks.

5 Prediction Model and Mix Design Methodology of Fly Ash–Slag Alkali-Activated Composites

5.1 PREAMBLE

An empirical relationship for predicting the compressive strength (MPa) of the fly ash–slag AAC has been developed on the basis of fly ash/slag ratio (by weight), alkali content ($Na_2O\%$), silica content ($SiO_2\%$), water content (water/binder ratio by weight), curing temperature (°C), and curing duration (hours). Data corresponding to a thermal curing temperature of 85°C have been considered.

5.2 DEVELOPMENT OF EMPIRICAL RELATIONSHIPS FOR PREDICTING COMPRESSIVE STRENGTH OF FLY ASH–SLAG AACS

5.2.1 EMPIRICAL RELATIONSHIP BETWEEN ALKALI CONTENT AND COMPRESSIVE STRENGTH

The compressive strength has been plotted against the corresponding alkali content (A_c) as shown in Figure 5.1. The compressive strength of the specimens having fly ash/slag ratio = 70/30, water/binder ratio (w) = 0.38, different alkali contents (4% to 12%), and constant silica content (S_c) of 10% has been considered. Quadratic polynomial trend lines of the curves were obtained having the highest regression coefficient ($R^2 = 0.9946$), and the corresponding equation between alkali content and compressive strength (C_s) has been obtained as given below.

$$C_s = C_1. A_c^2 + C_2. A_c + C_3$$

where $C_1 = -1.18$, $C_2 = 20.15$, and $C_3 = -50.6$.

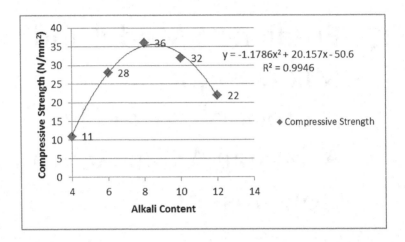

FIGURE 5.1   Regression analysis plot between alkali content and compressive strength.

5.2.2 Empirical Relationship between Silica Content and Compressive Strength

The compressive strength has been plotted against the corresponding silica content (S_c) as shown in Figure 5.2. The compressive strength of the specimens having fly ash/slag ratio = 70/30, water/binder ratio (w) = 0.38, different silica contents (4% to 14%), and constant alkali content (A_c) of 8% has been plotted. The quadratic polynomial trend line of the curve was obtained having regression coefficient $R^2 = 0.991$, and the corresponding equation between silica content (S_c) and compressive strength (C_s) is as given below.

$$C_s = C_4. S_c^2 + C_5. S_c + C_6$$

where $C_4 = -95$, $C_5 = 19.117$, and $C_6 = -52.203$.

5.2.3 Empirical Relationship between Compressive Strength and Water Content

The compressive strength has been plotted against the water/binder ratio (w) as shown in Figure 5.3. The compressive strength of the specimens having fly ash/slag ratio = 70/30, constant silica contents (S_c) 8%, and constant alkali content (A_c) of 8% and different water/binder ratios (w) has been plotted. Quadratic polynomial trend lines of the curves were obtained having the highest regression coefficient ($R^2 = 0.994$), and the corresponding equation between water/binder ratio and compressive strength (C_s) is as given below.

$$C_s = C_7. w^2 + C_8. w + C_9$$

where $C_7 = -35.239$, $C_8 = 193.46$, and $C_9 = -229.49$.

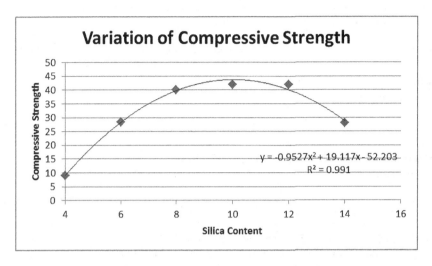

FIGURE 5.2   Regression analysis plot between silica content and compressive strength.

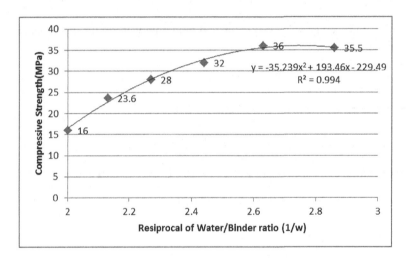

FIGURE 5.3   Regression analysis plot between water content and compressive strength.

5.2.4 PREDICTION OF COMPRESSIVE STRENGTH OF THE ALKALI-ACTIVATED SLAG COMPOSITE

The relationship between compressive strength (C_s), alkali content (A_c), silica content (S_c), and water/binder ratio (w) may be represented as follows:

$$C_s \; \alpha \; A_c. \, S_c \, / \, w$$

$$C_s = \eta \, A_c. \, S_c \, / \, w \tag{5.1}$$

where η = Constant of proportionality.

From Figure 5.2

$$C_s = C_1. A_c^2 + C_2. A_c + C_3 \tag{5.2}$$

where $C_1 = -1.18$, $C_2 = 20.15$, and $C_3 = -50.6$.
    From Figure 5.2

$$C_s = C_4. S_c^2 + C_5. S_c + C_6 \tag{5.3}$$

where $C_4 = -95$, $C_5 = 19.117$, and $C_6 = -52.203$.
    From Figure 5.2

$$C_s = C_7. w^2 + C_8. w + C_9 \tag{5.4}$$

where $C_7 = -35.239$, $C_8 = 193.46$, and $C_9 = -229.49$
    Combining regression Equations 5.2, 5.3, and 5.4 in Equation 5.1

$$C_s = \eta \left[\left(C_1. A_c^2 + C_2. A_c + C_3 \right) \left(C_4. S_c^2 + C_5. S_c + C_6 \right) / \left(C_7. w^2 + C_8. w + C_9 \right) \right] \tag{5.5}$$

Assuming the average value maximum and minimum value of the water/binder ratio considered

$$w = (0.35 + 0.5)/2 = 0.425$$

Substituting the value of 'w' in Equation 5.4

$$C_s = 34.08 \, \text{MPa}$$

Substitution the values of $A_c = 8\%$, $S_c = 8\%$, $C_s = 34.08$ MPa in Equation 5.5

$$\eta = 1.207669$$

Representing Equation 5.5 as below

$$C_s = \eta \, \overline{C_s}$$

$$\overline{C_s} = \left[\left(C_1. A_c^2 + C_2. A_c + C_3 \right) \left(C_4. S_c^2 + C_5. S_c + C_6 \right) / \left(C_7. w^2 + C_8. w + C_9 \right) \right]$$

where
    $C_1 = -1.18$, $C_2 = 20.15$, $C_3 = -50.6$
    $C_4 = -95$, $C_5 = 19.117$, $C_6 = -52.203$
    $C_7 = -35.239$, $C_8 = 193.46$, and $C_9 = -229.49$

5.2.5 Relationship between Slag Content and Compressive Strength

The compressive strength of the specimens having constant silica contents (S_c) 8% and constant alkali content (A_c) of 8%, water/binder ratio (w) of 0.38, and different

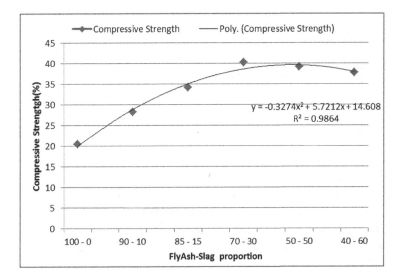

FIGURE 5.4 Regression analysis plot between fly ash–slag proportion and compressive strength.

TABLE 5.1
Multiplying Factors (K_{sl}) for Different Slag Contents (%)

Slag Content (%)	Multiplying Factor (Curing Temperature 85°C)
10%	0.7
15%	0.85
30%	1.0
50%	0.97
60%	0.94

fly ash/slag ratios has been considered and plotted. Data of thermal curing temperature 85°C have been considered.

A multiplying factor (K_{sl}) has been introduced to take care of slag content and is shown in Figure 5.4 and Table 5.1. Assuming $K_{sl} = 1$ for 30% slag content (fly ash/slag ratio = 70/30), K_{sl} for other slag contents has been obtained using the relationship obtained in Figure 5.4. (data for curing temperature 85°C have been considered).

5.2.6 MULTIPLYING FACTORS FOR CURING TEMPERATURE AND CURING DURATION

Again, another multiplying factor (K_{ct}) has been introduced to take care of curing temperature and is shown in Figure 5.5 and Table 5.2. The following values of K_{ct} may be considered.

The compressive strength of the specimens having constant silica contents (S_c) 8% and constant alkali content (A_c) of 8%, water/binder ratio (w) of 0.38, thermal curing temperature 85°C and 60°C, and different fly ash/slag ratios has been considered and

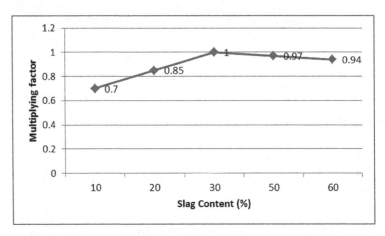

FIGURE 5.5 Multiplying factor (K_{sl}) vs. slag content. (considering thermal curing temperature 85°C).

TABLE 5.2
Multiplying Factors (K_{ct}) for Different Curing Temperatures and Slag Contents

Slag Content (%)	For 85°C	For 60°C
10%	1	0.93
15%	1	1.209
30%	1	1.27
50%	1	1.33
60%	1	1.32

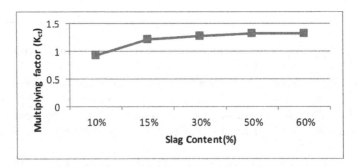

FIGURE 5.6 Multiplying factor (K_{ct}) for a curing temperature of 60°C. * Multiplying factor (K_{ct}) for curing temperature 85°C = 1 for all slag contents.

following constants are obtained (data available in 3.6.3.5.1.1 (Chapter 3) have been considered).

Again, another multiplying factor (K_{cd}) has been introduced to take care of curing duration and is shown in Figure 5.6 and Table 5.3. The following values of K_{cd} are provided below (data available in section 3.6.3.5.1.1 (Chapter 3) have been considered).

TABLE 5.3
Multiplying Factor (K_{cd}) for Different Curing Durations and Slag Contents

Slag Content (%)	K_{cd} for Different Thermal Curing Durations (for a Curing Temperature of 85°C)		
	24 Hrs	48 Hrs	72 Hrs
10%	0.9	1	0.95
20%	0.89	1	0.92
30%	0.85	1	0.91
50%	0.92	1	0.82
60%	0.92	1	0.76

Slag Content (%)	K_{cd} for Different Thermal Curing Durations (for a Curing Temperature of 60°C)		
	24 Hrs	48 Hrs	72 Hrs
10%	0.9	1	1.05
20%	0.89	1	1.07
30%	0.86	1	1.08
50%	0.82	1	1.11
60%	0.81	1	1.12

5.2.7 MULTIPLYING FACTORS FOR DIFFERENT FINENESS OF SLAG

The effect of fineness of slag on compressive strength for different fineness of slag (fineness of fly ash is around 150 μ) is presented graphically in Figures 5.7 and 5.8. The multiplying factor (K_f) for different fineness of slag has been calculated from Table 5.4.

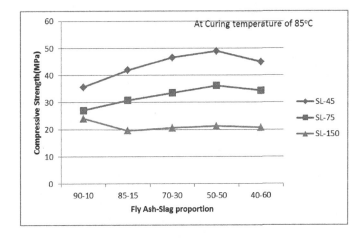

FIGURE 5.7 Compressive strength for different fineness and slag content at a curing temperature of 85°C.

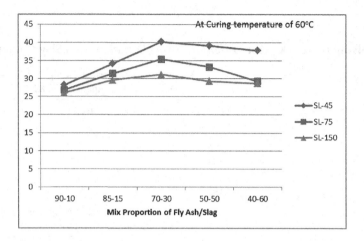

FIGURE 5.8 Compressive strength for different fineness and slag content at a curing temperature of 60°C.

TABLE 5.4
Multiplying Factor (K_f) for Different Fineness

Slag Content (%)	Temperature Degree Centigrade	Multiplying Factors (K_f) According to Fineness		
		Fineness (micron)		
		<45	45–75	75–150
10	85	1	0.95	0.92
20	85	1	0.92	0.87
30	85	1	0.88	0.77
50	85	1	0.85	0.75
60	85	1	0.77	0.76
10	60	1	0.76	0.68
20	60	1	0.74	0.47
30	60	1	0.72	0.44
50	60	1	0.74	0.43
60	60	1	0.76	0.46

5.3 PREDICTION MODEL

The empirical relationships have been presented in earlier sections separately and in combined form. Multiplying factors are also introduced. Finally, considering all the aspects, the compressive strength prediction model has been proposed as follows:

$$C_s = K_{sl} . K_{ct} . K_{cd} . K_f . \eta . \overline{C_s} \qquad (5.6)$$

where

$\overline{C_s} = [(C_1 . A_c^2 + C_2 . A_c + C_3)(C_4 . S_c^2 + C_5 . S_c + C_6)/(C_7 . w^2 + C_8 . w + C_9)]$

$C_1 = -1.18$, $C_2 = 20.15$, $C_3 = -50.6$, $C_4 = -95$, $C_5 = 19.117$, $C_6 = -52.203$

$C_7 = -35.239$, $C_8 = 193.46$, $C_9 = -229.4$

K_{sl} = Multiplying factor for taking the effect of slag content from Table 5.1

K_{ct} = Multiplying factor for taking the effect of curing temperature from Table 5.2.

K_{cd} = Multiplying factor for taking the effect of curing duration from Table 5.3

K_f = Multiplying factor for taking the effect of fineness of slag from Table 5.4

η = Constant of proportionality = 1.207669 (obtained earlier)

A_c = Alkali content (% Na_2O) as defined earlier

S_c = Silica content (% SiO_2) as defined earlier

W = Water–Binder ratio as defined earlier

5.3.1 VALIDATION OF THE PREDICTION MODEL

The calculated compressive strengths using the proposed prediction model have been compared with a different set of experimental results and are presented in Table 5.5 for validation.

TABLE 5.5

Comparison Between Predicted and Experimental Values of Compressive Strength

Fly Ash/Slag Proportion	Ac (%)	Sc (%)	W/B ratio	Curing Temperature (Degree C)	Curing Duration (Hrs)	Fineness of Slag (micron)	$\bar{C}_s$	K_{sl}	K_{ct}	K_{cd}	K_f	Predicted Value of Compressive Strength (MPa)	Actual Compressive Strength (MPa)	Percentage Deviation
90-10	4	8	0.35	85	48	45 - 75	12.4463	0.7	1	1	1	10.53	10.34	1.80
90-10	6	8	0.35	85	48	45 - 75	31.1382	0.7	1	1	1	26.35	24.81	5.84
90-10	8	8	0.35	85	48	45 - 75	39.2642	0.7	1	1	1	33.23	30.68	7.67
90-10	10	8	0.35	85	48	45 - 75	36.8242	0.7	1	1	1	31.16	29.06	6.74
85-15	4	8	0.35	85	48	45 - 75	12.4463	0.85	1	1	1	12.79	12.11	5.32
85-15	6	8	0.35	85	48	45 - 75	31.1382	0.85	1	1	1	32	30.16	5.75
85-15	8	8	0.35	85	48	45 - 75	39.2642	0.85	1	1	1	40.35	38.33	5.01
85-15	10	8	0.35	85	48	45 - 75	36.8242	0.85	1	1	1	37.84	39.68	-4.86
70-30	4	8	0.35	85	48	45 - 75	12.4463	1	1	1	1	15.05	14.34	4.72
70-30	6	8	0.35	85	48	45 - 75	31.1382	1	1	1	1	37.65	35.33	6.16
70-30	8	8	0.35	85	48	45 - 75	39.2642	1	1	1	1	47.47	45.51	4.13
70-30	10	8	0.35	85	48	45 - 75	36.8242	1	1	1	1	44.52	45.57	-2.36
50-50	4	8	0.35	85	48	45 - 75	12.4463	0.97	1	1	1	14.6	15.27	-4.59
50-50	6	8	0.35	85	48	45 - 75	31.1382	0.97	1	1	1	36.52	34.57	5.34
50-50	8	8	0.35	85	48	45 - 75	39.2642	0.97	1	1	1	46.05	43.7	5.10
50-50	10	8	0.35	85	48	45 - 75	36.8242	0.97	1	1	1	43.18	40.26	6.76
40-60	4	8	0.35	85	48	45 - 75	12.4463	0.94	1	1	1	14.14	14.97	-5.87
40-60	6	8	0.35	85	48	45 - 75	31.1382	0.94	1	1	1	35.39	33.75	4.63
40-60	8	8	0.35	85	48	45 - 75	39.2642	0.94	1	1	1	44.62	42.68	4.35
40-60	10	8	0.35	85	48	45 - 75	36.8242	0.94	1	1	1	41.85	39.75	5.02
90-10	8	4	0.35	85	48	45 - 75	8.91317	0.7	1	1	1	7.54	7.19	4.64
90-10	8	6	0.35	85	48	45 - 75	27.825	0.7	1	1	1	23.55	22.3	5.31
90-10	8	8	0.35	85	48	45 - 75	39.2642	0.7	1	1	1	33.23	31.51	5.18
90-10	8	10	0.35	85	48	45 - 75	43.2306	0.7	1	1	1	36.59	34.51	5.68

(Continued)

TABLE 5.5 (*Continued*)
Comparison Between Predicted and Experimental Values of Compressive Strength

Fly Ash/Slag Proportion	Ac (%)	Sc (%)	W/B ratio	Curing Temperature (Degree C)	Curing Duration (Hrs)	Fineness of Slag (micron)	$\bar{C}_s$	K_{sl}	K_{ct}	K_{cd}	K_f	Predicted Value of Compressive Strength (MPa)	Actual Compressive Strength (MPa)	Percentage Deviation
90-10	8	12	0.35	85	48	45 - 75	39.7243	0.7	1	1	1	33.62	32.23	4.13
90-10	8	14	0.35	85	48	45 - 75	28.7454	0.7	1	1	1	24.33	23.15	4.85
85-15	8	4	0.35	85	48	45 - 75	8.91317	0.85	1	1	1	9.16	8.65	5.57
85-15	8	6	0.35	85	48	45 - 75	27.825	0.85	1	1	1	28.59	26.99	5.60
85-15	8	8	0.35	85	48	45 - 75	39.2642	0.85	1	1	1	40.35	38.26	5.18
85-15	8	10	0.35	85	48	45 - 75	43.2306	0.85	1	1	1	44.43	41.69	6.17
85-15	8	12	0.35	85	48	45 - 75	39.7243	0.85	1	1	1	40.82	38.47	5.76
85-15	8	14	0.35	85	48	45 - 75	28.7454	0.85	1	1	1	29.54	27.97	5.31
70-30	8	4	0.35	85	48	45 - 75	8.91317	1	1	1	1	10.78	10.22	5.19
70-30	8	6	0.35	85	48	45 - 75	27.825	1	1	1	1	33.64	34.54	-2.68
70-30	8	8	0.35	85	48	45 - 75	39.2642	1	1	1	1	47.47	45.31	4.55
70-30	8	10	0.35	85	48	45 - 75	43.2306	1	1	1	1	52.27	54.08	-3.46
70-30	8	12	0.35	85	48	45 - 75	39.7243	1	1	1	1	48.03	46.07	4.08
70-30	8	14	0.35	85	48	45 - 75	28.7454	1	1	1	1	34.75	36.09	-3.86
50-50	8	4	0.35	85	48	45 - 75	8.91317	0.97	1	1	1	10.45	9.9	5.26
50-50	8	6	0.35	85	48	45 - 75	27.825	0.97	1	1	1	32.63	31.53	3.37
50-50	8	8	0.35	85	48	45 - 75	39.2642	0.97	1	1	1	46.05	43.44	5.67
50-50	8	10	0.35	85	48	45 - 75	43.2306	0.97	1	1	1	50.7	52.06	-2.68
50-50	8	12	0.35	85	48	45 - 75	39.7243	0.97	1	1	1	46.59	44.86	3.71
50-50	8	14	0.35	85	48	45 - 75	28.7454	0.97	1	1	1	33.71	32.8	2.70
40-60	8	4	0.35	85	48	45 - 75	8.91317	0.94	1	1	1	10.13	9.49	6.32
40-60	8	6	0.35	85	48	45 - 75	27.825	0.94	1	1	1	31.62	29.45	6.86
40-60	8	8	0.35	85	48	45 - 75	39.2642	0.94	1	1	1	44.62	41.84	6.23
40-60	8	10	0.35	85	48	45 - 75	43.2306	0.94	1	1	1	49.13	45.96	6.45
40-60	8	12	0.35	85	48	45 - 75	39.7243	0.94	1	1	1	45.15	41.89	7.22
40-60	8	14	0.35	85	48	45 - 75	28.7454	0.94	1	1	1	32.67	31.52	3.52

5.4 MIX DESIGN METHOD

5.4.1 Parameters in Mix Proportioning

i. Quality of fly ash, ground granulated blast furnace slag, and its fineness
ii. Composition of activator solution
iii. Total quantity of water
iv. Total quantity of aggregates
v. Curing conditions

The chemical composition and particle size distribution of fly ash and slag must be known prior to AAC mix design. X-ray fluorescence (XRF) analysis may be performed to determine the chemical composition of fly ash and slag. The following criteria may be followed for fly ash and slag for good performance of fly ash–slag AACs.

5.4.1.1 Fly Ash

a. Particle size of fly ash should be less than 45 microns.
b. $SiO_2 + Al_2O_3 + Fe_2O_3$ should be minimum 70%.
c. Total SiO_2 should be minimum 35%.
d. Loss on ignition should be less than 5%.
e. Insoluble residue should be less than 4%.
f. MgO content should be less than 5%.
g. MnO content should be less than 5.5%.
h. Total sulfur as sulfur trioxide should be less than 3%

5.4.1.2 Slag

i. Particle size of fly ash and slag should be less than 45 microns.
j. Glass content should be minimum 85%.
k. Loss on ignition should be less than 5%.
l. Insoluble residues should be less than 4%.
m. MgO content should be less than 17%.
n. MnO content should be less than 5.5%.
o. Sulfide sulfur should be less than 2%.
p. The $[(CaO + MgO + Al_2O_3)/SiO_2]$ ratio in slag should be minimum 1.0.
q. The CaO/SiO_2 ratio in slag should be between 0.5 and 2.0.
r. SiO_2/Al_2O_3 ratio in slag = 1.6 to 3.0.
s. To ensure good hydration properties, the hydration modulus (HM)

$$HM = \left[\left(CaO + MgO + Al_2O_3\right)/SiO_2\right] \text{ should exceed } 1.4.$$

5.4.1.3 Composition of Activator Solution

The quantity of polymeric gel formation depends on the composition and quantity of activator solution. In the present guidelines, the alkaline activator liquid is the combination of sodium silicate and sodium hydroxide pellets. It may be noted here that when sodium silicate was added to sodium hydroxide, the compressive strength was found higher. The addition of sodium silicate tends to increase the degree of polycondensation in the polymerization process, while the addition of sodium hydroxide may increase the degree of dissolution. The alkali content and silica content of the mix are expressed in terms of percentage of Na_2O and SiO_2 (by weight of source material).

5.4.1.4 Water Content

Total water content in the mix includes water from sodium hydroxide, sodium silicate, and extra water added to an activator solution. The water to source material (Fly ash + Slag) ratio of 0.35 to 0.41 is found to be suitable for moderate workability and reasonable strength.

5.4.1.5 Curing Conditions

Curing temperature may be kept between 60°C and 85°C, depending on the fly ash–slag proportion. If the slag content is within 30%, the curing temperature may be kept at 60°C otherwise may be kept at 85°C. In the presence of slag, water curing may be performed.

5.4.1.6 Procedure for Mix Design

The following procedure may be adopted for mix design of alkali-activated paste.

Step 1: The chemical analysis of fly ash and slag may be determined using XRF analysis.
Step 2: Check the suitability of fly ash and slag. It should meet the requirement presented earlier in Section 5.4.1.
Step 3: The proportion of fly ash to slag may be taken as 70:30.
Step 4: Determine % (Na_2O) by weight of (fly ash + slag) for target strength to be achieved by assuming % SiO_2 constant or determine the % SiO_2 by assuming % Na_2O constant [Figures 5.1 and 5.2].
Step 5: Calculate the percentage of SiO_2 based on the silicate ratio of the mix, i.e., SiO_2/Na_2O ratio between 1.25 and 1.5.
Step 6: Calculate the quantity of sodium silicate solution for a desired percentage of SiO_2 from Figure 5.9.

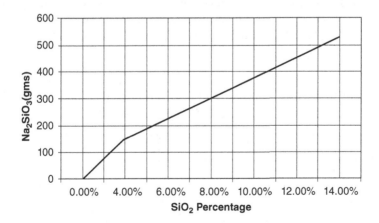

FIGURE 5.9 Quantity of sodium silicate solution (in gm) for different percentages of SiO_2 per 1000 gm of source material, i.e. (fly ash + slag)

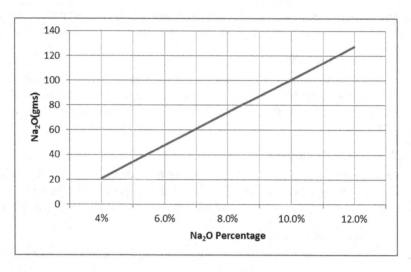

FIGURE 5.10 Quantity of NaOH solids (in gm) for different percentages of Na_2O per 1000 gm of source material, i.e., (fly ash + slag)

Step 7: Find out Na_2O available from sodium hydroxide and sodium silicate solution, respectively.

Step 8: Calculate the quantity of NaOH solids for the desired percentage of Na_2O from Figure 5.10.

Step 9: Calculate the quantity of water available in sodium hydroxide and sodium silicate solution.

Step 10: Calculate the additional quantity of water required to obtain the water to source material (fly ash + slag) ratio between 0.33 and 0.41 from Tables 5.6 and 5.7. Adjust the water to achieve required workability/flow by trial.

Step 11: Determine the initial setting time of alkali-activated paste using standard Vicat's apparatus at room temperature. If setting time is less than 30 minutes, then increase the quantity of water and check the setting time again.

TABLE 5.6

Extra Water to be Added into the AAC Mix Based on the Water/Binder Ratio and $SiO_2\%$ in Activator Solution ($Na_2O = 8\%$ Constant)

Water/Binder Ratio	SiO_2 Percentage					
	4%	6%	8%	10%	12%	14%
0.35	231.41	183.73	136.05	88.37	40.69	NR
0.38	261.41	213.73	166.05	118.37	70.69	23.01
0.41	291.41	243.73	196.05	148.37	58.18	53.01

Note: NR – Extra water is not required in the activator solution.

TABLE 5.7

Extra Water to be Added into the AAC Mix Based on the Water/Binder Ratio and $Na_2O\%$ in Activator Solution (SiO_2 = 10% Constant)

Water/Fly Ash–Slag Ratio	Na_2O Percentage				
	4%	6%	8%	10%	12%
0.35	99.98	94.18	88.37	82.56	76.76
0.38	129.98	124.18	118.37	112.56	106.76
0.41	159.98	154.18	148.37	142.56	136.76

Step 12: Prepare trial mortar mix using sand (confirms zone III as per IS 383) to source material ratio by weight equal to 1.0.

Step 13: Check the workability of the alkali-activated mortar mix using a mini flow table test as per ASTM C1437-07 with a modification, that is, the table was raised and dropped 15 times in about 15 seconds. The reason for modification is that some mix may tend to spread more than the diameter of the table (250 mm) and the purpose of the test would be lost. The flow diameter may be between 160 and 180 mm to achieve moderate workability. Adjust the quantity of water to achieve the desired workability.

Step 14: Keep fly ash–slag specimens undisturbed for four hours in a mold and then keep them in an oven for 48 hrs. After thermal curing, the specimens may be kept in air for seven days.

Step 15: Determine compressive strength of test specimen at the age of seven days.

Step 16: Adjust the quantity of sand for getting the desired strength of the mix with a maximum 5% standard deviation and decide the final mix proportion and repeat steps 11 to 14.

5.4.2 MIX DESIGN EXAMPLE

Mix proportion is given for AACs using the following data, for a target strength of 35 MPa after 48 hours of oven curing followed by five days of ambient air curing.

Data given:

A. Chemical composition of slag as determined using XRF analysis is shown in Table 5.8:

Fineness of fly ash and slag preferably less than 45 microns.

Alkaline activator:

a. Composition of sodium hydroxide pellets with 97% purity with specific gravity of 2.15.

b. Sodium silicate solution ($Na_2O = 8\%$, $SiO_2 = 26.5\%$ and 65.50% water) with silicate modulus ~ 3.3 and bulk density of 1410 kg/m³.

Step 1: Refer Table 5.9: Check the suitability of fly ash and slag

TABLE 5.8
Chemical Composition of Fly Ash and Slag by Obtained XRF

Oxide (wt.%)	SiO$_2$	Al$_2$O$_3$	Fe$_2$O$_3$	CaO	MgO	SO$_3$	Na$_2$O	K$_2$O	LOI*	Others
Fly ash	65.81	22.17	3.23	1.24	1.01	0.47	0.16	2.62	1.57	1.72
Slag	37.25	10.24	1.1	42.17	3.82	2.13	0.19	0.66	0.81	1.63

a Loss on ignition

TABLE 5.9
Check the Suitability of Fly Ash and Slag

Sl.no.	Properties of Fly Ash	Values Obtained	Desired Range	Remarks
1	$SiO_2 + Al_2O_3 + Fe_2O_3$	91.21%	Minimum = 70%	O.K.
2	SiO_2	65.81	35%	O.K.
3	SiO_2/Al_2O_3 ratio	2.97	1.6–3.0	O.K.
4	% on un-burnt carbon	1.57%	Less than 5%	O.K.
5	% MgO	1.01%	Less than 17%	O.K.
6	% SO$_3$	0.47%	Less than 3%	O.K.
7	% Chloride	0.0%	Less than .05%	O.K.

Sl.no.	Properties of Slag	Values Obtained	Desired Range	Remarks
1	$[(CaO + MgO + Al_2O_3) / SiO_2]$	1.51	Minimum = 1	O.K.
2	CaO/SiO_2 ratio	1.13	0.5–2.0	O.K.
3	SiO_2/Al_2O_3 ratio	3.63	1.6–3.0	O.K.
4	Hydration modulus (HM) HM = $[(CaO + MgO + Al_2O_3) / SiO_2]$	1.85	>1.4	O.K.
5	% on un-burnt carbon	0.81%	Less than 5%	O.K.
6	% MgO	3.82%	Less than 17%	O.K.
7	Glass content	89.66%	Minimum 85%	O.K.
8	Particles finer than 45 microns	100%	Minimum 100%	O.K.

Remark: Fly Ash and Slag (GGBS) satisfies physical and chemical requirements

Step 2: Mix proportioning

Guideline - $Na_2O = 6\%$ to 8% and $\% \ SiO_2 = 10\%$ to 12% may be considered.

Referring to target grade (M35) (Figures 5.1 and 5.2), let us assume the $Na_2O = 8\%$ on the higher side and value of $SiO_2 = 10\%$ on the lower side. Assuming water/binder ratio = 0.38.

Step 3: Based on the above data, calculating the quantities of sodium hydroxide pellets and sodium silicate solution and water required to make the paste moderately workable (flow between 160 and 180 mm). Let us consider curing temperature 85°C.

Let us assume the concentration of sodium silicate solution in terms of percentage of $SiO_2 = 12\%$.

From Figure 5.9

Quantity of sodium silicate solution = 452.83 gm per kg
of source material (fly ash + slag).

Let us assume the sodium hydroxide pellets having percentage of $Na_2O = 8\%$
From Figure 5.10

Quantity of sodium hydroxide pellets = 58.23 gm per kg of source material.

Determine extra water required for $8\% \ Na_2O$, $12\% \ SiO_2$ and

Let Water – binder ratio = 0.38

From Table 5.4.1 a and 5.4.1 b.

Extra water required in activator solution = 70.29 gm / kg of soured material

It can be now summarized as

1. Fly ash + slag = 1000 gm
2. Sodium silicate = 452.83 gm
3. Sodium hydroxide pellets = 58.23 gm
4. Extra water = 70.29 gm

Step 4: Proportioning of mortar mix

Let us consider sand to (fly ash + slag) ratio of 1:1.
Determine the flow diameter of mortar mix.
The flow diameter of the mortar using a mini flow table test as per ASTM C 1437-07 (modified) is found equal to 180 mm.
So, the mix confirms to moderate workability.

Hence O.K.

Casted 12 specimens of size 50 mm × 50 mm × 50 mm using the above mix proportions.

Specimens are oven-cured at 85 ± 2°C for 48 hours and then left in ambient air for five days.

The average compressive strength obtained at 7 days = 40.4 MPa > 35 MPa. Hence O.K.

The mix proportion may be accepted.

References

1. Davidovits, J. "Geopolymer cements to minimize carbon dioxide greenhouse warming." *Ceramic Transactions* 37 (1993): 165–182.
2. Van Deventer, J. S. J., J. L. Provis, and P. Duxson. "Technical and commercial progress in the adoption of geopolymer cement." *Minerals Engineering* 29 (2012): 89–104.
3. van Deventer, J. S. J., J. L. Provis, P. Duxson, and D. G. Brice. "Chemical research and climate change as drivers in the commercial adoption of alkali activated materials." *Waste and Biomass Valorization* 1, no. 1 (2010): 145–155.
4. Bakharev, T., J. G. Sanjayan, and Y.-B. Cheng. "Alkali activation of Australian slag cements." *Cement and Concrete Research* 29, no. 1 (1999): 113–120.
5. van Jaarsveld, J. G. S. and J. S. J. Van Deventer. "Effect of the alkali metal activator on the properties of fly ash-based geopolymers." *Industrial & Engineering Chemistry Research* 38, no. 10 (1999): 3932–3941.
6. Roy, D. M. "Alkali-activated cements opportunities and challenges." *Cement and Concrete Research* 29, no. 2 (1999): 249–254.
7. Palomo, A., M. W. Grutzeck, and M. T. Blanco. "Alkali-activated fly ashes: A cement for the future." *Cement and Concrete Research* 29, no. 8 (1999): 1323–1329.
8. Gruskovnjak, A., B. Lothenbach, L. Holzer, R. Figi, and F. Winnefeld. "Hydration of alkali-activated slag: Comparison with ordinary Portland cement." *Advances in Cement Research* 18, no. 3 (2006): 119–128.
9. Sindhunata, J. L. Provis, G. C. Lukey, H. Xu, and J. S. J. van Deventer. "Structural evolution of fly ash based geopolymers in alkaline environments." *Industrial & Engineering Chemistry Research* 47, no. 9 (2008): 2991–2999.
10. Hardjito, D., S. E. Wallah, D. M. J. Sumajouw, and B. Vijaya Rangan. "On the development of fly ash-based geopolymer concrete." *Materials Journal* 101, no. 6 (2004): 467–472.
11. Woolard, C. D., K. Petrus, and M. Van Der Horst. "The use of a modified fly ash as an adsorbent for lead." *Water SA* 26 (2000): 531. ISSN: 0378-4738.
12. Baldwin, G., P. E. Rushbrook, and C. G. Dent. "The testing of hazardous waste to assess their suitability for landfill disposal." Harwell Report, AERE-R10737, November 1982.
13. Shirin, S. and A. Jamal. "Neutralization of acidic mine water using flyash and overburden." *Rasayan Journal of Chemistry* 11, no. 1 (2018): 74–79.
14. Akbar, H., G. Krishan, S. D. Prajapati, and R. Saini. "Determination of reactive silica (SiO_2) of fly ash." *Rasayan Journal of Chemistry* 9, no. 1 (2016): 27–30.
15. Provis, J. L., R. J. Myers, C. E. White, V. Rose, and J. S. J. Van Deventer. "X-ray microtomography shows pore structure and tortuosity in alkali-activated binders." *Cement and Concrete Research* 42, no. 6 (2012): 855–864.
16. Mahmood, A. H., S. J. Foster, and A. Castel. "Development of high-density geopolymer concrete with steel furnace slag aggregate for coastal protection structures." *Construction and Building Materials* 248 (2020): 118681.
17. Divya Krishnan K., P. T. Ravichandran, and V. K. Gandhimathi. "Experimental study on properties of concrete using ground granulated blast furnace slag and copper slag as a partial replacement for cement and fine aggregate." *Rasayan Journal of Chemistry* 10, no. 2 (2017): 600–605.
18. Kavisri, M., P. Senthilkumar, M. S. Gurukumar, and K. J. Pushparaj. "Experimental study on effects of stabilization of clayey soil using copper slag and ggbs." *Rasayan Journal of Chemistry* 11, no. 1 (2018): 111–117.

19. Bakharev, T. "Geopolymeric materials prepared using Class F fly ash and elevated temperature curing." *Cement and Concrete Research* 35, no. 6 (2005): 1224–1232.

20. Chindaprasirt, P., T. Chareerat, and V. Sirivivatnanon. "Workability and strength of coarse high calcium fly ash geopolymer." *Cement and Concrete Composites* 29, no. 3 (2007): 224–229.

21. Ismail, I., S. A. Bernal, J. L. Provis, R. San Nicolas, S. Hamdan, and J. S. J. van Deventer. "Modification of phase evolution in alkali-activated blast furnace slag by the incorporation of fly ash." *Cement and Concrete Composites* 45 (2014): 125–135.

22. Oh, J. E., P. J. M. Monteiro, S. S. Jun, S. Choi, and S. M. Clark. "The evolution of strength and crystalline phases for alkali-activated ground blast furnace slag and fly ash-based geopolymers." *Cement and Concrete Research* 40, no. 2 (2010): 189–196.

23. Kumar, S., R. Kumar, and S. P. Mehrotra. "Influence of granulated blast furnace slag on the reaction, structure and properties of fly ash based geopolymer." *Journal of Materials Science* 45, no. 3 (2010): 607–615.

24. Lloyd, R. R. The Durability of Inorganic Polymer Cements. PhD thesis. University of Melbourne, 2008.

25. Bellum, R. R., K. Muniraj, and S. R. Madduru. "Investigation on modulus of elasticity of fly ash-ground granulated blast furnace slag blended geopolymer concrete." *Materials Today: Proceedings* 27 (2020 Jan 1): 718–723.

26. Puligilla, S. and P. Mondal. "Role of slag in microstructural development and hardening of fly ash-slag geopolymer." *Cement and Concrete Research* 43 (2013): 70–80.

27. Yang, T., X. Yao, Z. Zhang, and H. Wang. "Mechanical property and structure of alkali-activated fly ash and slag blends." *Journal of Sustainable Cement-Based Materials* 1, no. 4 (2012): 167–178.

28. Thokchom, S., P. Ghosh, and S. Ghosh. "Resistance of fly ash based geopolymer mortars in sulfuric acid." *ARPN Journal of Engineering and Applied Science* 4, no. 1 (2009): 65–70.

29. Thokchom, S., P. Ghosh, and S. Ghosh. "Durability of fly ash geopolymer mortars in nitric acid–effect of alkali (Na_2O) content." *Journal of Civil Engineering and Management* 17, no. 3 (2011): 393–399.

30. Kong, D. L. Y. and J. G. Sanjayan. "Effect of elevated temperatures on geopolymer paste, mortar and concrete." *Cement and Concrete Research* 40, no. 2 (2010): 334–339.

31. Gopalakrishnan, R. and K. Chinnaraju. "Durability of alumina silicate concrete based on slag/fly-ash blends against acid and chloride environments." Materiali in tehnologije/*Materials and Technology* 50, no. 6 (2016): 929–937.

32. Mehta, P. K. and P. J. M. Monteiro. *Concrete Microstructure, Properties, and Materials*, 3rd edn. McGraw-Hill, New York, 2006, ISBN: .

33. *U.S. Geological Survey*. Mineral Commodity Summaries 2018. U.S. Geological Survey, Reston, VA, 2018.

34. Hasanbeigi, A., L. Price, H. Lu, and W. Lan. "Analysis of energy-efficiency opportunities for the cement industry in Shandong Province, China: A case study of 16 cement plants." *Energy* 35 (2010): 3461–3473.

35. Valipour, M., M. Yekkalar; M. Shekarchi, and S. Panahi. "Environmental assessment of green concrete containing natural zeolite on the global warming index in marine environments." *Journal of Cleaner Production* 65 (2014): 418–423.

36. Rashad, A.M. and S. R. Zeedan. "The effect of activator concentration on the residual strength of alkali-activated fly ash pastes subjected to thermal load." *Construction and Building Materials* 25 (2011): 3098–3107.

37. Rashad, A. M. "An exploratory study on high-volume fly ash concrete incorporating silica fume subjected to thermal loads." *Journal of Cleaner Production* 87 (2015): 735–744.

38. Duxson, P., J. L. Provis, G. C. Lukey, and J. S. J. Van Deventer. "The role of inorganic polymer technology in the development of 'green concrete'." *Cement and Concrete Research* 37, no. 12 (2007): 1590–1597.

39. McLellan, B. C., R. P. Williams, J. Lay, A. Van Riessen, and G. D. Corder. "Costs and carbon emissions for geopolymer pastes in comparison to ordinary Portland cement." *Journal of Cleaner Production* 19, no. 9-10 (2011): 1080–1090.

40. Perera, D. S., O. Uchida, E. R. Vance, and K. S. Finnie. "Influence of curing schedule on the integrity of geopolymers." *Journal of Materials Science* 42, no. 9 (2007): 3099–3106.

41. Davidovits, J., *Geopolymer Chemistry and Applications*, Ch. 2, pp. 19–35. Geopolymer Institute, Saint-Quentin, France, 2008.

42. Duxson, P., G. C. Lukey, and J. S. J. van Deventer. "Physical evolution of Na- geopolymer derived from metakaolin up to 1000 C." *Journal of Materials Science* 42, no. 9 (2007): 3044–3054.

43. Dimas, D., I. Giannopoulou, and D. Panias. "Polymerization in sodium silicate solutions: A fundamental process in geopolymerization technology." *Journal of Materials Science* 44, no. 14 (2009): 3719–3730.

44. Vu, M. C., T. Satomi, H. Takahashi. Influence of initial water, moisture, and geopolymer content on geopolymer modified sludge. *Construction and Building Materials* 235 (2020 Feb 28): 117420.

45. Provis, J. L. and J. S. J. Van Deventer, eds. *Geopolymers: Structures, Processing, Properties and Industrial Applications*. Elsevier, 2009.

46. Criado, M., A. Fernández-Jiménez, A. Palomo, I. Sobrados, and J. Sanz. "Effect of the SiO_2/Na_2O ratio on the alkali activation of fly ash. Part II: 29Si MAS-NMR Survey." *Microporous and Mesoporous Materials* 109, no. 1-3 (2008): 525–534.

47. Shere, I. and A. Malani. "Polymerization kinetics of a multi-functional silica precursor studied using a novel Monte Carlo simulation technique." *Physical Chemistry Chemical Physics* 20, no. 5 (2018): 3554–3570.

48. Fernández-Jiménez, A., A. Palomo, I. Sobrados, and J. Sanz. "The role played by the reactive alumina content in the alkaline activation of fly ashes." *Microporous and Mesoporous Materials* 91, no. 1-3 (2006): 111–119.

49. Chen, X., A. Meawad, and L. J. Struble. "Method to stop geopolymer reaction." *Journal of the American Ceramic Society* 97, no. 10 (2014): 3270–3275.

50. Provis, J. L., V. Rose, S. A. Bernal, and J. S. J. van Deventer. "High- resolution nanoprobe X-ray fluorescence characterization of heterogeneous calcium and heavy metal distributions in alkali-activated fly ash." *Langmuir* 25, no. 19 (2009): 11897–11904.

51. Provis, J. L. and J. S. J. Van Deventer, eds. *Alkali Activated Materials: State-of-the-Art Report, RILEM TC 224-AAM*, Vol. 13. Springer Science & Business Media, Dordrecht, the Netherlands, 2013.

52. Criado, M., A. Fernández-Jiménez, and A. Palomo. "Alkali activation of fly ash: Effect of the SiO_2/Na_2O ratio: Part I: FTIR study." *Microporous and Mesoporous Materials* 106, no. 1-3 (2007): 180–191.

53. Nath, S. K., S. Mukherjee, S. Maitra, and S. Kumar. "Ambient and elevated temperature geopolymerization behaviour of class F fly ash." *Transactions of the Indian Ceramic Society* 73, no. 2 (2014): 126–132.

54. Van Deventer, J. S. J., J. L. Provis, P. Duxson, and G. C. Lukey. "Reaction mechanisms in the geopolymeric conversion of inorganic waste to useful products." *Journal of Hazardous Materials* 139, no. 3 (2007): 506–513.

55. Huanhai, Z., W. Xuequan, X. Zhongzi, and T. Mingshu. "Kinetic study on hydration of alkali-activated slag." *Cement and Concrete Research* 23, no. 6 (1993): 1253–1258.

56. Haha, M. B., G. Le Saout, F. Winnefeld, and B. Lothenbach. "Influence of activator type on hydration kinetics, hydrate assemblage and microstructural development of alkali activated blast-furnace slags." *Cement and Concrete Research* 41, no. 3 (2011): 301–310.

57. Gebregziabiher, B. S., R. J. Thomas, and S. Peethamparan. "Temperature and activator effect on early-age reaction kinetics of alkali-activated slag binders." *Construction and Building Materials* 113 (2016): 783–793.

58. Duxson, P., G. C. Lukey, F. Separovic, and J. S. Van Deventer. "Effect of alkali cations on aluminum incorporation in geopolymeric gels." *Industrial & Engineering Chemistry Research* 44, no. 4 (2005 Feb 16): 832-839.

59. Ma, Y., J. Hu, and G. Ye. "The effect of activating solution on the mechanical strength, reaction rate, mineralogy, and microstructure of alkali-activated fly ash." *Journal of Materials Science* 47, no. 11 (2012): 4568–4578.

60. Lothenbach, B. "Thermodynamic equilibrium calculations in cementitious systems." *Materials and Structures* 43, no. 10 (2010): 1413–1433.

61. Lothenbach, B., D. Damidot, T. Matschei, and J. Marchand. "Thermodynamic modelling: State of knowledge and challenges." *Advances in Cement Research* 22, no. 4 (2010): 211–223.

62. Provis, J. L. and S. A. Bernal. "Geopolymers and related alkali-activated materials." *Annual Review of Materials Research* 44 (2014): 299–327.

63. Ozçelik, V. O. and C. E. White. "Nanoscale charge-balancing mechanism in alkali-substituted calcium–silicate–hydrate gels." *The Journal of Physical Chemistry Letters* 7, no. 24 (2016): 5266–5272.

64. Hobbs, D. W. "Concrete deterioration: Causes, diagnosis, and minimising risk." *International Materials Reviews* 46, no. 3 (2001): 117–144.

65. Mundra, S., M. Criado, S. A. Bernal, and J. L. Provis. "Chloride-induced corrosion of steel rebars in simulated pore solutions of alkali-activated concretes." *Cement and Concrete Research* 100 (2017): 385–397.

66. Lee, W. K. W. and J. S. J. Van Deventer. "Structural reorganisation of class F fly ash in alkaline silicate solutions." *Colloids and Surfaces A: Physicochemical and Engineering Aspects* 211, no. 1 (2002): 49–66.

67. Zuo, Y., M. Nedeljković, and G. Ye. "Pore solution composition of alkali-activated slag/fly ash pastes." *Cement and Concrete Research* 115 (2019): 230–250.

68. Xiuren, S. C. L. Y. T. "Preliminary investigation on the activation mechanism of phosphorus slag." *Journal of Southeast University* 13, no. 3 (1989): 115–121.

69. Davidovits, J. "Geopolymers of the first generation: SILIFACE-process." *Geopolymer* 88, no. 1 (1988): 49–67.

70. Phair, J. W. and J. S. J. Van Deventer. "Effect of silicate activator pH on the leaching and material characteristics of waste-based inorganic polymers." *Minerals Engineering* 14, no. 3 (2001): 289–304.

71. Purdon, A. O. "The action of alkalis on blast furnace slag." *Journal of the Society of Chemical Industry* 59 (1940): 191–202.

72. Glukhovsky, V. D., Y. Zaitsev, and V. Pakhomov. "Slag-alkaline cements and concrete structure, properties, technological and economical aspects of use." *Journal of Silicate Industries* 10 (1983): 197–200.

73. Hunter, R.obert J. *Introduction to Mmodern Ccolloid Sscience,*. Vol. 7. Oxford University Press, Oxford, 1993.

74. Davidovits, J. "Geopolymer chemistry and properties." In *Proceedings of Ist International Conference on Geopolymer '88*, France, June 1–3, 1988, Vol. 1, pp. 25–48.

75. Lee, N. K. and H. K. Lee. "Reactivity and reaction products of alkali-activated, fly ash/slag paste." *Construction and Building Materials* 81 (2015 Apr 15): 303–312.

76. Provis, J. L. and J. S. J. Van Deventer. "Geopolymerisation kinetics. 1. In situ energy-dispersive X-ray diffractometry." *Chemical Engineering Science* 62, no. 9 (2007): 2309–2317.

77. Xu, H. and J. S. J. Van Deventer. "The geopolymerisation of alumino-silicate minerals." *International Journal of Mineral Processing* 59, no. 3 (2000): 247–266.

78. De Silva, Pre, K. Sagoe-Crenstil, and Vute Sirivivatnanon. "Kinetics of geopolymerization: role of Al_2O_3 and SiO_2." *Cement and Concrete Research* 37, no. 4 (2007): 512–518.

79. Rattanasak, U. and P. Chindaprasirt. "Influence of NaOH solution on the synthesis of fly ash geopolymer." *Minerals Engineering* 22, no. 12 (2009): 1073–1078.

80. Hajimohammadi, A., J. L. Provis, and J. S. J. van Deventer. "The effect of silica availability on the mechanism of geopolymerisation." *Cement and Concrete Research* 41, no. 3 (2011): 210–216.

81. Sindhunata, J. S. J. Van Deventer, G. C. Lukey, and H. Xu. "Effect of curing temperature and silicate concentration on fly-ash-based geopolymerization." *Industrial & Engineering Chemistry Research* 45, no. 10 (2006): 3559–3568.

82. Kumar, S. and R. Kumar. "Mechanical activation of fly ash: Effect on reaction, structure and properties of resulting geopolymer." *Ceramics International* 37, no. 2 (2011): 533–541.

83. Wang, S.-D. and K. L. Scrivener. "Hydration products of alkali activated slag cement." *Cement and Concrete Research* 25, no. 3 (1995): 561–571.

84. Shi, C. and R. L. Day. "Some factors affecting early hydration of alkali-slag cements." *Cement and Concrete Research* 26, no. 3 (1996): 439–447.

85. Ryu, G. S., Y. B. Lee, K. T. Koh, and Y. S. Chung. "The mechanical properties of fly ash-based geopolymer concrete with alkaline activators." *Construction and Building Materials* 47 (2013): 409–418.

86. Panias, D., I. P. Giannopoulou, and T. Perraki. "Effect of synthesis parameters on the mechanical properties of fly ash-based geopolymers." *Colloids and Surfaces A: Physicochemical and Engineering Aspects* 301, no. 1-3 (2007): 246–254.

87. Görhan, G. and G. Kürklü. "The influence of the NaOH solution on the properties of the fly ash-based geopolymer mortar cured at different temperatures." *Composites Part B: Engineering* 58 (2014): 371–377.

88. Duxson, P., J. L. Provis, G. C. Lukey, S. W. Mallicoat, W. M. Kriven, and J. S. J. Van Deventer. "Understanding the relationship between geopolymer composition, microstructure and mechanical properties." *Colloids and Surfaces A: Physicochemical and Engineering Aspects* 269, no. 1-3 (2005): 47–58.

89. Chindaprasirt, P., P. De Silva, K. Sagoe-Crentsil, and S. Hanjitsuwan. "Effect of SiO_2 and Al_2O_3 on the setting and hardening of high calcium fly ash-based geopolymer systems." *Journal of Materials Science* 47, no. 12 (2012): 4876–4883.

90. Thakur, R. N. and S. Ghosh. "Effect of mix composition on compressive strength and microstructure of fly ash based geopolymer composites." *ARPN Journal of Engineering and Applied Sciences* 4, no. 4 (2009): 68–74.

91. Nematollahi, B. and J. Sanjayan. "Effect of different superplasticizers and activator combinations on workability and strength of fly ash based geopolymer." *Materials & Design* 57 (2014): 667–672.

92. Qureshi, M. N. and S. Ghosh. "Effect of silicate content on the properties of alkali-activated blast furnace slag paste." *Arabian Journal for Science and Engineering* 39, no. 8 (2014): 5905–5916.

93. Chi, M. "Effects of dosage of alkali-activated solution and curing conditions on the properties and durability of alkali-activated slag concrete." *Construction and Building Materials* 35 (2012): 240–245.

94. Bernal, S. A., R. Mejía de Gutiérrez, A. L. Pedraza, J. L. Provis, E. D. Rodriguez, and S. Delvasto. "Effect of binder content on the performance of alkali-activated slag concretes." *Cement and Concrete Research* 41, no. 1 (2011): 1–8.

95. Živica, V. "Effects of type and dosage of alkaline activator and temperature on the properties of alkali-activated slag mixtures." *Construction and Building Materials* 21, no. 7 (2007): 1463–1469.

96. Bernal, S. A., R. S. Nicolas, J. S. J. van Deventer, and J. L. Provis. "Alkali-activated slag cements produced with a blended sodium carbonate/sodium silicate activator." *Advances in Cement Research* 28, no. 4 (2016): 262–273.

97. Aydın, S. and B. Baradan. "Effect of activator type and content on properties of alkali-activated slag mortars." *Composites Part B: Engineering* 57 (2014): 166–172.

98. Wang, W.-C., H.-Y. Wang, and M.-H. Lo. "The fresh and engineering properties of alkali activated slag as a function of fly ash replacement and alkali concentration." *Construction and Building Materials* 84 (2015): 224–229.

99. Phoo-ngernkham, T., A. Maegawa, N. Mishima, S. Hatanaka, and P. Chindaprasirt. "Effects of sodium hydroxide and sodium silicate solutions on compressive and shear bond strengths of FA–GBFS geopolymer." *Construction and Building Materials* 91 (2015): 1–8.

100. Marjanović, N., M. Komljenović, Z. Baščarević, V. Nikolić, and R. Petrović. "Physical–mechanical and microstructural properties of alkali-activated fly ash–blast furnace slag blends." *Ceramics International* 41, no. 1 (2015): 1421–1435.

101. Jang, J. G., N. K. Lee, and H.-K. Lee. "Fresh and hardened properties of alkali-activated fly ash/slag pastes with superplasticizers." *Construction and Building Materials* 50 (2014): 169–176.

102. Bernal, S. A., R. Mejía de Gutiérrez, and J. L. Provis. "Engineering and durability properties of concretes based on alkali-activated granulated blast furnace slag/metakaolin blends." *Construction and Building Materials* 33 (2012): 99–108.

103. Khater, H. M. "Studying the effect of thermal and acid exposure on alkali-activated slag geopolymer." *Advances in Cement Research* 26, no. 1 (2014): 1–9.

104. Lee, N. K. and H.-K. Lee. "Influence of the slag content on the chloride and sulfuric acid resistances of alkali-activated fly ash/slag paste." *Cement and Concrete Composites* 72 (2016): 168–179.

105. Komljenović, M., Z. Baščarević, N. Marjanović, and V. Nikolić. "External sulfate attack on alkali-activated slag." *Construction and Building Materials* 49 (2013): 31–39.

106. Bakharev, T., J. G. Sanjayan, and Y.-B. Cheng. "Sulfate attack on alkali-activated slag concrete." *Cement and Concrete Research* 32, no. 2 (2002): 211–216.

107. Guerrieri, M., J. Sanjayan, and F. Collins. "Residual strength properties of sodium silicate alkali activated slag paste exposed to elevated temperatures." *Materials and Structures* 43, no. 6 (2010): 765–773.

108. Rovnaník, P., P. Bayer, and P. Rovnaníková. "Characterization of alkali activated slag paste after exposure to high temperatures." *Construction and Building Materials* 47 (2013): 1479–1487.

109. Yip, Ch. K., G. C. Lukey, and J. S. J. Van Deventer. "The coexistence of geopolymeric gel and calcium silicate hydrate at the early stage of alkaline activation." *Cement and Concrete Research* 35, no. 9 (2005): 1688–1697.

110. Garcia-Lodeiro, I., A. Palomo, A. Fernández-Jiménez, and D. E. Macphee. "Compatibility studies between NASH and CASH gels. Study in the ternary diagram $Na_2O–CaO–Al_2O_3–SiO_2–H_2O$." *Cement and Concrete Research* 41, no. 9 (2011): 923–931.

111. García-Lodeiro, I., A. Fernández-Jiménez, A. Palomo, and D. E. Macphee. "Effect of calcium additions on N–A–S–H cementitious gels." *Journal of the American Ceramic Society* 93, no. 7 (2010): 1934–1940.

112. Komljenović, M., Z. Baščarević, and V. Bradić. "Mechanical and microstructural properties of alkali-activated fly ash geopolymers." *Journal of Hazardous Materials* 181, no. 1-3 (2010): 35–42.

113. Wardhono, A., D. W. Law, and A. Strano. "The strength of alkali-activated slag/fly ash mortar blends at ambient temperature." *Procedia Engineering* 125 (2015): 650–656.

114. Zhang, Z., J. L. Provis, A. Reid, and H. Wang. "Mechanical, thermal insulation, thermal resistance and acoustic absorption properties of geopolymer foam concrete." *Cement and Concrete Composites* 62 (2015): 97–105.

115. Collins, F. and J. G. Sanjayan. "Effect of pore size distribution on drying shrinking of alkali-activated slag concrete." *Cement and Concrete Research* 30, no. 9 (2000): 1401–1406.

116. Kong, D. L. Y., J. G. Sanjayan, and K. Sagoe-Crentsil. "Comparative performance of geopolymers made with metakaolin and fly ash after exposure to elevated temperatures." *Cement and Concrete Research* 37, no. 12 (2007): 1583–1589.

117. Ling, Y., K. Wang, X. Wang, and S. Hua. "Effects of mix design parameters on heat of geopolymerization, set time, and compressive strength of high calcium fly ash geopolymer." *Construction and Building Materials* 228 (2019): 116763.

118. Somna, K., C. Jaturapitakkul, P. Kajitvichyanukul, and P. Chindaprasirt. "NaOH-activated ground fly ash geopolymer cured at ambient temperature." *Fuel* 90, no. 6 (2011): 2118–2124.

119. Everett, D. H. *Basic Principles of Colloid Science*. Royal Society of Chemistry, London, 2007.

120. Zuhua, Z., Y. Xiao, Z. Huajun, and C. Yue. "Role of water in the synthesis of calcined kaolin-based geopolymer." *Applied Clay Science* 43, no. 2 (2009): 218–223.

121. Catalfamo, P., S. Di Pasquale, F. Corigliano, and L. Mavilia. "Influence of the calcium content on the coal fly ash features in some innovative applications." *Resources, Conservation and Recycling* 20, no. 2 (1997 Jun 1): 119–125.

122. Lee, W. K. W. and J. S. J. Van Deventer. "The effect of ionic contaminants on the early-age properties of alkali-activated fly ash-based cements." *Cement and Concrete Research* 32, no. 4 (2002): 577–584.

123. Hamidi, R. M., Z. Man and K. A. Azizi. " Concentration of NaOH and the effect on the properties of fly ash based geopolymer." In *4th International Conference on Process Engineering and Advanced Materials Procedia Engineering*, Kuala Lumpur, 2016, Vol. 148, pp. 189–193

124. Mikolajczyk, A., A. Gajewicz, B. Rasulev, N. Schaeublin, E. Maurer-Gardner, S. Hussain, J. Leszczynski, and T. Puzyn. "Zeta potential for metal oxide nanoparticles: A predictive model developed by a nano-quantitative structure–property relationship approach." *Chemistry of Materials* 27, no. 7 (2015): 2400–2407.

125. Revathi, T., R. Jeyalakshmi, N. P. Rajamane, and M. Sivasakthi. "Evaluation of the role of Cetyltrimethylammoniumbromide (CTAB) and Acetylenic Glycol (AG) admixture on fly ash based geopolymer." *Oriental Journal of Chemistry* 33, no. 2 (2017): 783–792.

126. Sivasakthi, M., R. Jeyalakshmi, N. P. Rajamane, and T. Revathi. "Use of analytical techniques for the identification of the geopolymer reactions." *Oriental Journal of Chemistry* 33, no. 4 (2017): 2103–2110.

127. Mužek, M. N., J. Zelić, and D. Jozić. "Microstructural characteristics of geopolymers based on alkali-activated fly ash." *Chemical and Biochemical Engineering Quarterly* 26, no. 2 (2012 Jul 4): 89–95.

128. Oderji, S. Y., B. Chen, M. R. Ahmad, and S. F. Shah. "Fresh and hardened properties of one-part fly ash-based geopolymer binders cured at room temperature: Effect of slag and alkali activators." *Journal of Cleaner Production* 225 (2019 Jul 10): 1-10.

129. Lee, W. K. W. and J. S. J. Van Deventer. "Use of infrared spectroscopy to study geopolymerization of heterogeneous amorphous aluminosilicates." *Langmuir* 19, no. 21 (2003): 8726–8734.

130. Duxson, P., J. L. Provis, G. C. Lukey, F. Separovic, and J. S. J. van Deventer. "29Si NMR study of structural ordering in aluminosilicate geopolymer gels." *Langmuir* 21, no. 7 (2005): 3028–3036.

131. Fernández-Jiménez, A., A. Palomo, and M. Criado. "Microstructure development of alkali-activated fly ash cement: A descriptive model." *Cement and Concrete Research* 35, no. 6 (2005): 1204–1209.

132. Palomo, Á., S. Alonso, A. Fernandez-Jiménez, I. Sobrados, and J. Sanz. "Alkaline activation of fly ashes: NMR study of the reaction products." *Journal of the American Ceramic Society* 87, no. 6 (2004): 1141–1145.

133. Xu, H. and J. S. J. van Deventer. "The effect of alkali metals on the formation of geopolymeric gels from alkali-feldspars." *Colloids and Surfaces A: Physicochemical and Engineering Aspects* 216, no. 1-3 (2003): 27–44.

134. Falcone, J. S. *Soluble Silicates*. American Chemical Society, Washington, DC, 1982.

135. Duxson, P. S. W. M., S. W. Mallicoat, G. C. Lukey, W. M. Kriven, and J. S. J. van Deventer. "The effect of alkali and Si/Al ratio on the development of mechanical properties of metakaolin-based geopolymers." *Colloids and Surfaces A: Physicochemical and Engineering Aspects* 292, no. 1 (2007): 8–20.

136. De Jong, B. H. W. S. and G. E. Brown Jr. "Polymerization of silicate and aluminate tetrahedra in glasses, melts, and aqueous solutions—I. Electronic structure of H6Si2O7, H6AlSiO71–, and H6Al2O72–." *Geochimica et Cosmochimica Acta* 44, no. 3 (1980): 491–511.

137. Qureshi, M. N. and S. Ghosh. "Effect of curing conditions on the compressive strength and microstructure of alkali-activated GGBS paste." *International Journal of Engineering Science Invention* 2, no. 2 (2013): 24–31.

Appendix

TYPICAL CALCULATION TO OBTAIN CONSTITUENTS OF AN ACTIVATOR SOLUTION

[A] Activator solution with sodium hydroxide pellets and sodium silicate solutionFor activator solution with 8% Na_2O and 10% SiO_2 by weight of source material (fly ash + slag)

a. Composition of sodium hydroxide (NaOH) pellets (as per test report)
   $Na_2O = 77.5\%$
   Water (H_2O) = 22.5%

b. Composition of sodium silicate (Na_2SiO_3) solution (as per test report)
   $Na_2O = 8\%$
   $SiO_2 = 26.50\%$
   Water (H_2O) = 65.5%

c. Required quantity of Na_2O and SiO_2 for 1000 g of (fly ash + slag)
   $Na_2O = 8\%$ of weight of (fly ash + slag)
   $$= \frac{8 \times 1000}{100} = 80 \text{ g}$$
   $SiO_2 = 10\%$ of weight of (fly ash + slag)
   $$= \frac{10 \times 1000}{100} = 100 \text{ g}$$

d. Quantity of Na_2SiO_3 required to get 100 g of SiO_2
   $$= \frac{100 \times 100}{26.50} = 377.36 \text{ g}$$

e. Quantity of Na_2O available in Na_2SiO_3
   $$= \frac{8.0 \times 377.36}{100} = 30.19 \text{ g}$$

f. Quantity of Na_2O required from NaOH pellets
   $= 49.81$ g

g. Quantity of NaOH pellets required
   $$= \frac{49.81 \times 100}{77.5} = 64.271 \text{ g}$$

h. Actual quantity of NaOH pellets required for 97% purity
   $= 64.271/0.97 = 66.26$ g

i. Water available in NaOH pellets and Na_2SiO_3 solution
   = Water available in (NaOH pellets + Na_2SiO_3 solution)
   $$= \frac{22.5 \times 66.26 + 65.50 \times 377.36}{100}$$
   $= 262.08$ g

j. Total water required for 1000 g of (fly ash + slag)
   Water to (fly ash + slag) ratio = 0.38
   Hence total quantity of water required = 0.38 × 1000 = 380 g

k. Extra water to be added
   = Total water required – available water in NaOH pellets and Na_2SiO_3 solution
   = 380 – 262.06
   = 117.92 g
l. Summary of activator solution constituents for 1000 g of (fly ash + slag)
   NaOH pellets = 66.26 g
   Na_2SiO_3 solution = 377.36 g
   Extra water = 117.92 g

Index

Printed in the United States
By Bookmasters